D0432633

the Reality Game

Samuel Woolley

the Reality Game

How the next wave of technology will break the truth and what we can do about it

ENDEAVOUR

An Hachette UK Company
www.hachette.co.uk

First published in Great Britain in 2020 by Endeavour,
an imprint of Octopus Publishing Group Ltd
Carmelite House
50 Victoria Embankment
London EC4Y 0DZ
www.octopusbooks.co.uk

First published in the US in 2020 by Public Affairs, an imprint of Perseus Books, LLC,
a subsidiary of Hachette Book Group, Inc.

ISBN 978-1-91306-812-7 (Hardback)
ISBN 978-1-91306-813-4 (Trade paperback)

A CIP catalogue record for this book is available from the British Library.

Printed and bound in the UK

1 3 5 7 9 10 8 6 4 2

For my mentor, Dr. Philip N. Howard,
who taught me what scholarship could be
while always creating space for me to be myself.

Contents

Author's Note

The concept of fake news burst onto the global scene in 2016 following the rise of blatantly false news stories and the flow of digital garbage during the presidential election in the United States. The specter of "fake news" was further fanned by suspicious rumors of smear campaigns against Russian athletes that arose during the summer Olympics in Rio de Janeiro and by misinformation about the Zika virus, which continued to spread in Brazil and elsewhere. The term "fake news" was quickly co-opted, though, by the powers that be. The very people who produced the junk content known by this moniker reclaimed the phrase as a means of undermining legitimate journalism, as a crutch to attack inconvenient scientific findings, or as a means to refute factual stories about their own misdeeds. The term "fake news" itself became a tool for spreading fake news.

With this in mind, I need to explain how I use a couple of terms and definitions that are important to the coming chapters and the arguments I make here. First, I try not to use the phrase "fake news." Instead, I use the term "misinformation," by which I mean the accidental spread of false content or "disinformation", by which I mean the purposeful spread of false content. I sometimes refer to "false news" or "junk news," and when I do I mean articles constructed to look like news that are not actually true, because they lack facts or verifiability. These types of articles, like the infamous pieces that came from the bogus *Denver Guardian* during the 2016 US election, are created with an intent to mislead, confuse, or, at times, make money (I will be covering this further in Chapter 3). I do not use "fake news" because the phrase has been repurposed as a tool to target articles and reports by actual journalists who write things with which thin-skinned politicians, litigious business executives, or incensed regular folks do not agree.

I refer to "computational propaganda" often. My colleagues and I originally came up with the term to refer to the use of automated tools (like Twitter bots) and algorithms over social media in attempts to manipulate public opinion. In this book I use the term more broadly to refer to the use of digital tools—from Facebook to augmented reality (AR) devices—to spread politically motivated information. Computational propaganda includes using social media to anonymously attack journalists in order to stop them from reporting. It includes leveraging digital voice systems designed to sound like humans to call voters over the phone and tell them lies about the opposition. It also includes using artificial intelligence (AI) and social bots—automated programs built to mimic people online—to fake human communication in order to trick the online algorithms that curate and prioritize our news.

Finally, I often talk about democracy and human rights. When I talk about "democracy," I am talking about democratic values: liberty, equality, justice, and so forth. I am not advocating for US-style democratic governance or for any other hybrid democratic-republican-parliamentary-presidential system. When I talk about "human rights," I have in mind the definition by the United Nations, which defines "human rights" as:

> *the rights inherent to all human beings, regardless of race, sex, nationality, ethnicity, language, religion, or any other status. Human rights include the right to life and liberty, freedom from slavery and torture, freedom of opinion and expression, the right to work and education, and many more. Everyone is entitled to these rights, without discrimination.*[1]

I argue that we should bake the values of democracy and human rights into our technology. We must prioritize equality and freedom in the tools we build so that the next wave of devices will not be used to further damage the truth.

My imagination makes me human and makes me a fool.

URSULA K. LE GUIN,
HARPER'S MAGAZINE (1990)

Chapter One

Truth Is Not T[illegible]

[illegible] Real, My Fake

[illegible] University. That [illegible]
[illegible], president of the Phil[illegible]
had just given his notice of the [illegible]
[illegible] recent research paper [illegible]
[illegible] that [illegible] the [illegible]
[illegible] propaganda [illegible]
[illegible] that the Filipino [illegible]
[illegible] social media army whose [illegible]
[illegible] I had admired [illegible]
[illegible] on this smoke[illegible]
[illegible] He denied, though [illegible]
[illegible] evidence [illegible]
the award-winning Filipino [illegible]
and editor of the publication [illegible]
[illegible] digital propaganda [illegible]
[illegible], like many other wo[illegible]
[illegible] manipulation [illegible]
I was the director of research [illegible]
[illegible] group, the Computational [illegible]
University of Oxford Internet Ins[illegible]
[illegible] of social media as a to[illegible]

Chapter One
Truth Is Not Technical

Your Real, My Fake

"Oxford University? That's a school for stupid people," said Rodrigo Duterte, president of the Philippines. It was July 24, 2017, and Duterte had just given his State of the Nation Address. A reporter had asked him about a recent research paper from Oxford University during a press conference following the event.[1] The paper in question detailed the social media propaganda expenses of various governments around the globe and claimed that the Filipino president spent approximately $200,000 for a social media army whose goal was to viciously defend him against critics.[2] Duterte admitted to the assembled crowd that he had, in fact, spent more than this amount for such purposes during his presidential campaign. He denied, though, that he continued to do so. He made this argument despite evidence to the contrary, cited in the Oxford paper, from the award-winning Filipino news outlet *Rappler*.[3] Maria Ressa, founder and editor of the publication, wrote that his regime continued to fund malicious digital propaganda and trolling campaigns against dissenters. Duterte, like many other world leaders, had turned social media into a tool for public manipulation.

I was the director of research for the Oxford team that drew Duterte's ire. Our group, the Computational Propaganda Project, was based at the university's Oxford Internet Institute. Our work was focused on explaining the use of social media as a tool for molding public opinion, hacking truth, and silencing protest. We detailed how automated Twitter "bot" profiles and trending algorithms were being used to influence people during pivotal political events. My colleagues and I wanted to uncover who was behind these underhanded campaigns and determine how they were spreading disinformation. More than anything, we wanted to know why they were doing what they were doing. What did they think they were achieving? It was not the first time we had struck a nerve with someone in a position of power through our research, but it was the first time a world leader had called us out specifically.

Soon after Duterte's attack on Oxford, *Rappler* produced a short video that explained how a variety of powerful political groups around the world, like the Duterte regime, used sites like Twitter, YouTube, and Facebook to troll their opposition (post deliberately offensive or incendiary online comments) and amplify spin campaigns.[4] The video said that these groups used bots and fake profiles "to create an alternative reality for people to believe in." Duterte's attack on Oxford, defaming the university and its research, was a parallel strategy for gaming the truth. He, like Narendra Modi in India, Donald Trump in the United States, and Jair Bolsonaro in Brazil, was combining ad hominem attacks, skewed logic, and social media tools to create a distorted version of what was real and what was fake.

The Next Wave of Technology and Disinformation

Though the past can tell us a great deal about what is to come, society must now pivot from concerns about digital "information operations" during past events and begin to look to the future. It is true that countries around the globe have experienced unprecedented levels of manipulation over social media in recent years. These changes to the way we communicate have weakened democracies and strengthened authoritarian regimes. Nevertheless, we need to take heed of something new on the horizon. The next wave of technology—from artificial intelligence (AI) to virtual reality

(VR)—will bring about a slew of newer and even more potent challenges to reality and the truth.

Although advances in artificial intelligence have created more effective methods for parsing data and prioritizing content for users on social media, they have also, and perhaps more concerningly, fundamentally changed how we spread information and who does the spreading. They have opened up an online world where the distinction between human and machine is increasingly blurry.

Manipulative social media advertisements during elections are certainly concerning, but what about political indoctrination in a virtual social media world? We cannot look away from this development, because advances in our digital tools are bringing about big changes to communication technology and society writ large. The next wave of technology will enable more potent ways of attacking reality than ever. In the humble words of Bachman-Turner Overdrive, "You ain't seen nothing yet."

For the better part of the last decade, I have been researching the ways in which propagandists leverage our technology and media systems. I have seen a rapid shift in how we perceive social media: once seen as exciting tools for connecting, communicating, and organizing, they are now often thought of as malicious platforms for spreading false news, political misinformation, and targeted harassment. And I am still witnessing efforts by some groups to control the messages we receive online. But every day I also learn about new initiatives and new technologies for pushing back and for prioritizing quality journalism, fact, and science over informational garbage.

In this book, I am going to tell you what I know. I'm going to talk through the recent history of political manipulation using digital tools, discuss how things look right now, and make educated guesses about what will come next. I'm also going to outline how we can respond and how we can reclaim our digital spaces. It's going to take work.

The "Assault" on Reality and the Truth

If you don't fund the State Department fully, then I need to buy more ammunition ultimately.

FORMER SECRETARY OF DEFENSE JAMES MATTIS

It took a few years of studying computational propaganda for me to come to a simple but important revelation: technology is what people make of it. In the spring of 2016, I was in Austin for the South by Southwest (SXSW) conference to give a talk on how social media can be used to game elections. After the presentation, I went out to a downtown bar near the conference center with some friends and colleagues. It was full of the odd mixture of people you get at an event like SXSW: techies, politicians, musicians, filmmakers, students, businesspeople, and so forth. At one point later in the night, and after several drinks, a man who had attended my talk came up to me. Sporting a pinstriped three-piece suit, very gelled hair, and lots of gold jewelry, he was dressed like a combination of a Wall Street banker and a member of the mob.

He told me that he was intrigued by my talk and had never heard of chatbots (automated profiles built to mimic real people) being used on social media to spread political content. He said that he worked in communications for "a government" and that he had just been tasked with taking over its social media operations. He was deliberately vague about all of this, and I never did find out where he was from, beyond somewhere in the "Indian Ocean region." I did learn that he had a proposition for me: Would I be interested in helping him build an army of bots to boost his government's image over social media? I laughed out loud. I had just given a talk about the perils of doing exactly this kind of thing, and this guy was almost guilelessly trying to get me to do precisely the opposite. Unsurprisingly, I emphatically told him no. We left it there and went our separate ways.

On another, very different, occasion, I was approached by a curator from the Victoria and Albert Museum, an art and design museum in London. He was putting together an exhibit on the future of design and wanted to know if I could build some kind of Twitter bot to go in it. The idea I landed on, along with another collaborator, was to build a socially oriented bot that would automatically share content on how its bot brethren could be used in politics and other social discussions online. It could also, to a degree, chat with people about politics and life. This bot, under the account name @futurepolitica1, would be transparent about its "botness" as it deliberately sought to educate people about the political misuse of technology.

The takeaway from these two separate stories is that a bot—or a VR program, a human-sounding "digital assistant," or a physical robot—can be built either to control channels of communication or to liberate those same channels. The tools that are already here, and those that are coming, can be harnessed for war or for peace, for propaganda or for art. How these tools are used depends on who is behind the digital wheel. Most democratic nations can agree upon absolutely unalienable human rights, but when it comes to how technology is used to manipulate, consensus is more difficult to reach. That is because the problems we face are not simply technical but social.

When I first started looking into how social media bots were being used to, say, defame activists online in the Middle East, it was easy for me to get hung up on the idea that these seemingly smart machines were automatically sending out cascades of harassment and spin. When I dug deeper, though, I realized that the vast majority of these campaigns were technologically rudimentary. The bots being used were simple to build, simple to launch, and simple in their communication. They repeated the same attacks and used the same hashtags over and over. The real problem was the people who launched the bots, and the people who paid for them. They were the conniving ones who came up with the idea of using bots to create the illusion of large-scale public online campaigns. It was humans who figured out that they could generate false hashtag trends on Twitter—there for everyone to see and click on via the site's sidebar—by using armies of bots to hugely boost the numbers of times a hashtag was shared.

Shifts in Technology, Shifts in Society

In 1991 the company Virtuality Group released the first networked and multi-player VR system for public use, the Virtuality 1000 series. Users experienced the platform through a bulky stereoscopic helmet and handheld joysticks, and a handful of arcades offered the public playtime with the new system. Systems for home use cost up to a whopping $73,000—just shy of $140,000 in today's dollars.[5] In the handful of decades since, VR has become much more accessible. Today you can pick up a Samsung Gear VR headset, which pairs with Samsung smartphones, for around $50. Yesterday's VR experiences offered blocky, low-resolution, simulation games like *Dactyl Nightmare*—a multi-level game not unlike the original *Donkey Kong*. Today's VR is plugged into budding social networks through apps like *Facebook Spaces* that are immersive and much more realistic. And VR is now being used for political and indoctrination purposes. Governments around the world are even beginning to use these systems to "train" ideal citizens.[6]

It is an understatement to say that things are changing, technologically and socially. The political bots and social media advertising campaigns that propagandists and political campaigns used to clobber reality during the 2016 US presidential campaign are becoming more sophisticated. They still require human guidance to be effective, but they are becoming steadily more automated—and more powerful. If we don't adapt to these changes, we run the risk of the global public completely losing trust in any information they encounter online.

Some researchers and pundits have suggested that social media and the internet have become the latest tools of war, that Facebook, YouTube, and Twitter have been weaponized by the powerful.[7] They argue that countries now use these digital weapons to attack one another in a battle of likes, retweets, and comments and that whoever wins on the virality front wins the day. It's true that groups in positions of power—militaries and governments among them—now use online communication platforms to spread propaganda and attack their opposition. Examples of these tactics abound, including, of course, the Russian influence campaign in the 2016 US election. But this isn't the whole story.

No media tool, from a book to a virtual simulation, is a weapon in and of itself. Social media are not actual weapons, and they aren't just used in information warfare. Widespread social problems created by national and global spikes in polarization and nationalism are primarily that—social. Online efforts to dupe people into donating money to scammers or false news campaigns designed to make money through clicks and views are economically driven. Campaigns to sway people's votes by using Twitter to falsely make a politician or idea seem more popular are political.

If we think of computational propaganda and other misuses of social media and technology simply as warfare, then we will fail to effectively address other underlying and complex issues. It is a combination of social, economic, and political problems that spurs manipulative uses of social media in the first place. There is more going on here than just the desire to do battle; this is more than simply a fight between those with access to troops and tanks. To solve the underlying issues we must not think in terms of defense and offense, but rather in terms of diplomacy and human rights. We must acknowledge that what we face is a broad and deep societal issue as well as one driven by new technology.

Reddit, Gab, Periscope, WhatsApp, WeChat, KakaoTalk, Instagram—all of these sites or applications, and hundreds of others like them, are social networking services or social media. Virtual reality and augmented reality are, similarly, immersive media tools. All of these function as communication technologies. They are vessels for spreading information.

The idea that any of these technologies, or any of the artificial intelligence or machine-learning (ML) capabilities that underpin them, can be weaponized exaggerates fear about pieces of code while overlooking the human role in uses of technology for purposes neutral, good, or evil.

Tools are not sentient—they do not act on their own. There is always a person behind a Twitter bot, a designer behind a VR game. A bot is just a way of automating and scaling what a human does online. Social media websites were designed by the Mark Zuckerbergs and Jack Dorseys of the world in order to connect people and, in so doing, make money. Many people, and not just their creators, thought that these new platforms would be phenomenal tools for advancing democracy. They would allow activists in Egypt to communicate about a revolution against an authoritarian regime. They would facilitate organization between journalists breaking a story on global rings of corruption. But—and here lies the failure of these platforms and those who are supposed to regulate them—they could also be used to control people, to harass them, and to silence them.

It is not just governments that have figured this out. Well-resourced actors, including politicos and corporations, special interest groups, intelligence agencies, and wealthy individuals, also use social media in attempts to manipulate not only what we read, see, hear, and watch online but also how we feel and what we believe. It is undoubtedly people with access to lots of money, time, and know-how who use social media most successfully to influence politics and social life. They're also the ones who are best able to manipulate the variety of emergent technologies, from deepfake videos to deep learning (DL), for their own selfish means and ends. But regular people and small far-right and far-left political groups have also figured out how to game trends on Twitter and control conversations on Facebook to achieve their own goals. There has been an opening up of who can sway public opinion and how they can do it.

We need to act now to prevent the misuse of tomorrow's technology. This book walks through the past, present, and future of how computer-

and internet-based tools are used to undermine reality and the truth. There are lots of stories in here about how we got to where we are, but there are also many stories about things that aren't yet in the news, that have not yet provoked a congressional hearing. There is also serious discussion about the potential problems posed by the use of new and future tools—alongside proposed solutions to these problems.

This book does not paint a doom-and-gloom picture of our technological world. It isn't a treatise on how technology companies screwed up or on how the addiction to social media of one particularly egotistical politician changed history. I talk about these things, but I focus much more on a variety of new media technologies and what we can do to ensure that they are used to build up the tenets of democracy rather than undermine them. This book takes an informed and cautiously optimistic approach to addressing the problems at hand. The truth is not broken yet. But the next wave of technology will break the truth if we do not act.

We live in a time when the quest to control reality has become something of a game, one mired in the ability to exploit the latest communication technologies in efforts to prioritize one notion of reality over another. That game is mostly played by the political elite and by disproportionately vocal extremist groups. We do not, however, have to play by their rules.

From Propaganda to Computational Propaganda

Jamal Khashoggi, the journalist murdered in late 2018 under extremely suspicious circumstances in the Saudi Arabian consulate in Istanbul, lived through the shift from the old world of propaganda to the new technological era of bending social reality. He, like other reporters around the world, saw Twitter and other social media networks become arenas for spreading the latest news and information. He and his colleagues also eventually realized that these tools were simultaneously being co-opted by governments—including Saudi Arabia—for their own Machiavellian purposes.

Khashoggi, publicly a cautious critic of Saudi policies, left his home country after experiencing a spate of harassment online and offline. Before leaving, he had been banned by the Saudi royal family from writing publicly or making media appearances.[8] The government there, like many other governments around the globe, still worked to exert control over all forms of media, but the Saudi government had also broadened its propaganda horizons. Khashoggi was also told not to use Twitter. In exile, once he had taken up a position as a columnist for the *Washington Post*, he defied that directive. But his personal and professional life online, and consequently aspects of his offline life, became untenable. According to the *New York Times*, Khashoggi experienced an orchestrated and tireless social media trolling campaign in the months leading up to his murder.[9] A team of Saudi "image makers" worked to defame and attack the journalist at every turn.

The trolls acted, according to the *Times*, at the behest of Saudi crown prince Mohammed bin Salman. Thousands of posts on Twitter targeted Khashoggi and his closest colleagues with vitriol and threats while simultaneously building up the Saudi government. Just before he was beaten and strangled to death, Khashoggi's online life had—by all accounts—become a living hell. He could not log onto Twitter without being barraged with disinformation, harassment, and hate. After the journalist's death, a similarly planned propaganda campaign worked to contradict allegations that the crown prince had ordered the killing. Armies of both bot-driven automated Twitter profiles and human-led accounts were instrumental in defaming and tearing down someone who, according to his friends and colleagues, was a tireless and fair-minded journalist.

The rise of digital disinformation and online political harassment—what I call "computational propaganda," Facebook calls "information operations," and most people call "fake news"—is a new way to manipulate people by using automated online tools and tactics.[10] It is used to target journalists, like Khashoggi, but it's also used to target politicians, public figures, and the general public. During the 2016 US election, numerous such online attacks, originating from both Russia and inside the United States, were used in attempts to manipulate the American people. Similar campaigns have been conducted around the world, orchestrated by world leaders and fringe political groups, from Duterte's troll machine in the Philippines to bin Salman's image polishers in Saudi Arabia.

While powerful political groups, from governments to militaries, still run the best-resourced and most pervasive campaigns, others have begun to adopt computational propaganda in their own amplification and suppression efforts. Even that outspoken person we all know on platforms like Twitter, Instagram, or Facebook can pay an illicit bot builder on a website such as Fiverr to get 1,000 or 10,000 automated accounts to amplify their rants about current events. But even as the political noise on social media becomes unbearable, things are changing. The tactics of computational propaganda are progressing and new tools are emerging. Trolling campaigns and botnets (groups of bots) are becoming more subtle and harder to track. Politicos are now beginning to seize upon advances in artificial intelligence to leverage the already widening rifts in society for political gain. They deploy smart technology to do the dirty work of campaigns: AI-doctored videos, increasingly individualized online political advertising campaigns, and facial recognition technology are among the tools used for these ends.

Propaganda, in and of itself, is certainly nothing new. The idea of manipulating how people think—and what people think about—has been around since at least ancient Greece.[11] The Greek origin myths and the legacy of the gods—of Zeus sitting atop Mount Olympus dictating weather patterns and striking down wayward mortals—were used to make grand political claims and lend legitimacy to dynasties.[12] In more recent conflicts, and during many elections in contemporary history, propaganda has played a key role in molding behavior and belief. The Cold War spurred a unique and memorable barrage of both Soviet and US propaganda.[13] Airborne leaflet propaganda—dropping purposefully crafted information on unsuspecting crowds from planes—is a form of psychological propaganda that originated as far back as World War I and continues to be employed today over war-torn regimes (in Syria, for instance).[14]

In some ways we are experiencing Cold War propaganda strategies today, amplified by powerful technology. But it's important to underscore the aspects of computational propaganda that are distinct from the propaganda of yesterday. What began as warfare tactics have become the political communication methods of the guy next door. Most obviously, this new version of manipulative information can be automated and is often completely anonymous.

Armies of political bots have been used to spread disinformation and political harassment for over a decade now. They trick trending algorithms on social media platforms, which are usually in charge of determining what news is prioritized. Suddenly the algorithm thinks that something that has the support of thousands of bots has the support of thousands of people, and it puts a link to that hashtag, video, or topic on the homepage of the platform. These automatons have also become a key tool for powerful politicos—government employees, cyber-troops, candidates for office—to magnify discord among the opposition, confuse people about how, when, and where to vote, and further polarize communities already facing a wide partisan divide. And regular people now use bots too. Political bots are, in many ways, the technological precursors of what is to come in the realm of digital political manipulation.

Human-mimicking bots and the rumors they were used to spread generated confusion in the wake of the Boston Marathon bombing in 2013.[15] They continue to be key tools for stymying democratic activism in Syria and were integral to the disinformation strategy deployed online against Jamal Khashoggi.[16] While the rest of the world asked pointed questions about Khashoggi's death, thousands of fake and automated accounts appeared on Twitter to extol the virtues of "the Kingdom" and insist that the Saudi government was not involved. The goal of these sorts of campaigns is to change how people think and feel about politics—not just to get them to vote for a certain candidate or take a different perspective about a news story, but to confuse, polarize, and disenchant.

The Role of Future Tech

A teenager sits alone at a computer desk. She is sixteen years old. She wears a VR headset and haptic gloves (that can simulate the sense of touch) and is logged onto the latest VR social media platform, a completely immersive experience in which you can meet anyone and do anything. This particular virtual world is a simulated hybrid of Twitter and Facebook but with more adventure, more engagement, and better costumes. As on these legacy social media sites, the teenage girl can use various features of the VR platform to access all sorts of information. She can interact with the latest stories on the entertainment industry, experiencing the happenings as if she were there alongside the celebrity or appearing in a TV show. She can also access educational modules and the news to learn about what is going on in her local area and around the world in a similarly engrossing way.

She can also get bullied and harassed on the platform, as she can at school or on social media, but here it happens in an unfettered virtual environment. As a soon-to-be voter, she can be targeted with false news reports and bogus information on elections and major political events. Now that she and her peers aren't simply reading stories and watching videos of events, now that they are immersed in an environment that enchants multiple senses, they can become unwitting victims of virtual disinformation. White supremacists, political extremists, and all manner of other predators can construct worlds within worlds in this social VR system where they can indoctrinate this young girl and others like her. They constantly barrage these kids, and even their parents and grandparents, with subtle political advertisements of dubious provenance and fake stories, such as the ones about how vaccines cause autism and other ailments.

This particular world and these circumstances do not exist yet, though several new and emergent social VR platforms are now available. With VR and other novel media tools, the issues of an unregulated and unconsidered digital sphere become all the more potent. These tools are coming. If we do not take action, we could very well end up with scenarios just like this. Digital propaganda is not just biased information, enhanced by automation and bots, that can be read on Facebook group pages or in YouTube comment sections. It is technologically enhanced propaganda that people can see, hear, and feel. In the not so distant future, it could be politically motivated information that is also tasted and smelled. This new way of spreading disinformation moves well beyond fake accounts on Twitter.

"Deepfakes"—videos so convincingly doctored that the eye cannot tell they're fake—are already being made to show politicians and public figures doing and saying things they haven't. And lying politicians can use the rise of these altered videos to claim that they were framed. They can deny the wrongdoing recorded in an actual video, insisting that they never made that gaffe or took that bribe. The video is a deepfake, they will argue. Automated voice calling systems, which sound just like a real person—with pauses, tics, and everything—have already been launched by Google, which bills its Duplex tool as an AI personal assistant. What happens when that system is used to call your grandmother to talk politics, or to threaten journalists over the phone? Imagine the possibilities for using such a system for political robocalls, or for push-polling—a technique sometimes used by campaigns to manipulate voters over the phone under the guise of an opinion poll. Virtual reality and augmented reality are immersive technologies that obscure the border between the physical and digital worlds and are useful for more than just entertainment and education. What happens when groups begin using VR as a means to manipulate?

More sophisticated chatbots are likely to supplant their more rudimentary social bot cousins. Whereas the bots used in the 2016 US election and the 2018 Mexican election were blunt instruments used to inflate likes, shares, and follows, AI-enabled chatbots will be able to convince real users through conversation. Passively using a bot to share a biased news article on vaccination with a group of people is a fairly unsophisticated way of changing minds. Deploying AI chatbots that are indistinguishable from people, can engage in real arguments, and can more effectively mimic emotion is likely to be far more effective. Beyond this, what will VR social bots look like? Will politicians and other groups be able to build groups of "smart" avatars to do their bidding—spreading their messages and attacking their enemies—in the virtual world?

To understand where technologically enhanced disinformation is heading, we have to look to the past. The next chapter details a short history of how social media websites and applications have been used for manipulation. Although for many people the 2016 US election was the first time that they experienced the impact of false news, it was not the first time that social media was gamed for political purposes. Between the founding of Facebook, YouTube, Twitter, and other major web 2.0 sites—the internet of social media—and 2016, these media tools were harnessed for coercion and control in lots of ways in many countries around the world. And the social media companies were aware of these events. They simply failed to act to curb computational propaganda before it got out of hand.

One of the main reasons I wanted to write this book was to empower people. I believe that the more people there are who understand the problem, the better. If we are educated about the history of dialogue around how the latest gadgets often go from tools for saving the world to implements for breaking democracy, then we can contextualize the current wave of propaganda within the larger history. The more we learn about computational propaganda and its elements, from false news to political trolling, the more we can do to stop it from taking hold. Today's propagandists, criminals, and con artists rely on people not understanding how technology and propaganda campaigns work in order to deceive them.

The more we know about these tactics, the less effective they are. And the more people there are who advocate for sensible solutions to stop the spread of junk news and unfair data-gathering practices, the better off we all will be.

Chapter Two
Breaking the Truth: Past, Present, and Future

Beginnings

"Are you with the CIA?" asked Jascha, a self-proclaimed anarchist software engineer.

"No," I said. "I'm a graduate student, but a lot of people get the two confused."

It was 2014. I was in a dark and messy basement in Budapest, Hungary, filled with computers and other technological odds and ends, with my research collaborator, whom I will call Motoko here, and five or six other people. We'd emailed back and forth with this group, a hacking collective, and had finally been allowed to visit them. At the time, Motoko, a few other researchers, and I were doing research fellowships at the Center for Media, Data, and Society at Central European University. We were part of a small team studying how social media bots were being used to target people online during political events. We thought that the hackers might know a thing or two about where to look.

Jascha told me that we were lucky, because he had just returned from the Euromaidan protests in Ukraine. He told us about all kinds of crazy stuff that had been going on there, both online and offline. It turned out that Jascha was a gold mine of information, though it took me a while to realize just how valuable he was. At first I thought he was just a foul-mouthed coder with some serious anarchist tendencies—someone whose opinions I should filter through a fine sieve. I was partly right about him, but as time went by some of his assertions and predictions turned out to be incredibly on point.

Jascha had gone to Ukraine in late 2013 to help anti-Russian protesters organize themselves technologically. He taught impromptu courses in encryption and cybersecurity, showed young Ukrainians how to code using Python, and then showed them how to leverage this knowledge for basic hacking. Bouncing from one squat or crowded apartment to another, Jascha was armed with his two laptops and assorted tech gear as his weapons of dissent. He witnessed many new ways of using manipulative, usually poor-quality information over social media to target regular people in the country. Later it would become clear that Ukraine was the frontier of computational propaganda. Now, when we want to understand where the future of fake news and political bots is going, we use Ukraine as a case study.

Among the many people he encountered during his travels, Jascha said, with a sidelong glance my way, were American spies trying to infiltrate the Ukrainian hacking collectives. We thought he would tell us about what was going on in Hungary, how Viktor Orbán and his government were putting restrictions on the media, the online sphere, and freedom of speech in general. He and his buddies did talk briefly about Hungary, but more importantly for us, they said that they hadn't noticed social media being used there to spread propaganda in any organized way. They were emphatic that we look instead at events in Eastern Europe and beyond. Ukraine and Poland, they told us, were the places to look to see both the positive and negative uses of social media and computing technology. Similar informational struggles were taking place in Turkey, the Middle East, and Mexico, they said. In other words, digital political manipulation was all around us, even then.

Where Does Digital Disinformation Come From?

When I ask most people about when they first started noticing widespread disinformation online, they talk about Ukraine. They bring up the slew of fake stories that came out after the alleged Russian downing of Malaysian Airlines Flight 17.[1] Sometimes they talk about the launch of Twitter bot armies to spread deceptive political content during the elections and crises that have occurred in Turkey, Mexico, or Syria. Some argue that the wave of fake news bots used on Twitter during the UK's Brexit campaigns represented the earliest moment in online false news. Others are sure that online disinformation originated with ISIS and other international terror groups, though most of the extremists who use social media for recruitment and messaging seem to have learned these online skills from more mainstream groups. Even more people are sure that it all began in Russia—and they aren't half wrong. Many of the underlying tactics of disinformation have roots in the Kremlin and date back to old-school media manipulation during the Cold War. But computational propaganda has broader international roots because it has evolved on an internet largely absent of borders.

What most people don't know is that one of the first well-documented uses of computational propaganda occurred during an event in the United States. It happened back in 2010, during the Massachusetts special Senate election between Scott Brown and Martha Coakley.[2] The race was a contentious one for a state that has long been a Democratic stronghold. The seat had been held for nearly fifty years by Ted Kennedy, the scion of that eminent Democratic family. Midway through the campaign, two computer science researchers at Wesleyan University noticed that a group of suspicious-looking Twitter accounts were launching what looked like coordinated attacks on Coakley.[3] The aggressors alleged that Coakley was anti-Catholic, a serious allegation in a state where nearly half the population self-identify as members of the Catholic Church.[4]

Upon closer inspection, the researchers noticed discrepancies in the accounts being used to defame the Democratic nominee. Most of the account profiles had no profile pictures or, when they did, they were stock images. Most of the users lacked biographical data and had very few followers. In fact, the accounts were mostly following one another—or following random combinations of disconnected accounts that hadn't followed them back. They were tweeting anti-Coakley content much more often or much more formulaically than the average Twitter user. Some sent out messages every few seconds and seemed to be posting on a time schedule. It turned out that the smear campaign was driven by bots. Automated accounts built to look like real residents of Massachusetts were being used to wage an "astroturf" (fake grassroots) battle against Coakley. The Wesleyan researchers traced the accounts back to a small group of Tea Party activists in Iowa. The digitally savvy partisan activists were using automated profiles to simultaneously attack the opposition while amplifying their candidate.

The group that launched the automated accounts was successful in many ways. Outlets from the *National Catholic Register* to the *National Review* reported on Coakley's supposed anti-Catholic tendencies.[5] The news pieces, some even citing the cascade of messages on Twitter as evidence for growing anti-Coakley sentiment, wrote about how people in Massachusetts were upset with the candidate over alleged discrimination against members of the church. Suddenly, the Democratic Party had a manufactured controversy on its hands.[6] Bots had given the allegations against Coakley the illusion of legitimacy and popularity. Eventually, the Republican Party won, taking a Senate seat that most pundits had considered staunchly blue.

Dialing Up

In 2013, a few years after the Brown–Coakley botnet debacle, I started as a PhD student at the University of Washington. During the Obama–Romney presidential election in 2012, I decided to study technology and political communication. I wanted to know how campaigns were using digital tools to connect with voters. After reading about the Obama digital team and the evolution of their data-oriented organizing apparatus, I felt like I was seeing the future of politics. The campaign was using huge collections of data on undecided voters to contact these voters and attempt to win them over. This early example of a political campaign making use of the tools of big data analysis to target individual voters was very different from the deceptive individual online ad targeting that groups like Cambridge Analytica peddled to Ted Cruz and Donald Trump in 2016.

It was clear that future campaigns would have to make use of similar data and technology-centric strategies if they wanted to be competitive in national politics. Without these tools and tactics, it would be impossible for them to keep up with the personalization of political marketing. In the Obama–Romney race, big data sets detailing information on citizens' behaviors and demographic data were being parsed using previously unavailable computational power. At that time, almost everyone was taken with the democratic potential of these amazing technologies. Those who criticized online political mobilization focused on "slacktivism"—that is, on what they saw as the half-baked organizing efforts of internet activists.[7] They focused much less of their attention on the underlying problems with the technological infrastructure, antidemocratic digital propaganda, the social media companies, and the governments that should have regulated them.[8]

When I moved to Seattle, political junkies and tech wonks were still reeling with the implications of two significant events: the Arab Spring of 2010–2011 and the Occupy movement, which began in the fall of 2011. Though different in focus and participants, both movements represented a large-scale grassroots effort that made use of digital technology for communication and organization against what each saw as systems of control. Researchers Alexandra Segerberg and Lance Bennett termed Occupy's strategy "connective action."[9] They wrote that the Occupy movement was not conceived as a highly organized and well-resourced collective action campaign but rather made use of a different brand of internet-driven endeavor "based on personalized content sharing across media networks." This was not slacktivism. They argued that Occupy represented something new, exciting, and unique.

Philip Howard, director of the Oxford Internet Institute and a professor at Oxford University, made similar observations about the use of the web before and during the Arab Spring. He argued that the broad array of connected information and communication technologies (ICTs) had spurred new forms of democratic engagement in some countries while cementing authoritarian perspectives and practices in others.[10] It was through Phil that I was first introduced to the idea of the social media bot.

While doing fieldwork in North Africa and the Middle East during the Arab Spring, Phil had heard about, and subsequently encountered in online research, armies of fake automated social media accounts built to look like real users. The botnets would barrage the Twitter hashtags used by democratic organizers with spam or malicious content so that these groups were unable to use Twitter to organize public meetings or communicate about the topics at hand. Propagandists used bots to amplify links to fake news stories or to suddenly and exponentially boost the follower numbers of online accounts associated with embattled leaders across the region. Artificially enhanced Twitter follower counts might not have kept some of the leaders in power, but these numbers did give the false impression that they had far more public support online than they did in reality.

Phil and I collaborated on early work studying the people who made and built these bots, and that became my dissertation topic. There was a lot to learn from the people who created social bot technology. Bots on platforms like Twitter were being used not only to spread political disinformation but also to provide unique scaffolding for civic engagement and for art and cultural critique. More generally, these "political" bots—as we began calling them—used automation in unique ways to communicate with both real people and other bots. They, and their builders, were operating at the edges of work and innovation in social media, automation, and artificial intelligence. Many of the early bots we examined were fairly simple, but I certainly saw a future where these digital automatons could be trained to operate more independently.

My colleagues and I had stumbled upon an emergent field of study—computational propaganda. We had to build an understanding of this emergent issue as we went along because there was very little work on social media and propaganda at the time. Our goal was to plot out the size and scale of the problem and to chart how bots and other tools were being used to manipulate public opinion around the world. We took what we knew, including work on the technical aspects of automated politicking as plotted out by a few groups of innovative computer scientists, and connected it to the larger sociological context. We built a database of knowledge that spoke to both the supply side of computational propaganda (who was building it and why) and the demand side (who was consuming it and why).

In the first few months we found case after case, in country after country, of social media bots being used to interrupt and alter political conversations. Once we realized that this was happening on such a global scale, we began to argue that the use of automation and algorithms over social media in attempts to manipulate public opinion was one of the most pressing problems facing democracy. I spent a lot of time learning as I went along back then, through trial and error. It was exciting, but also humbling. Some experts laughed when I told them that social bots could be used to manipulate political discussion online. As Phil and I put it early on, most people and companies thought of "bots as a nuisance to be detected and managed," not as a global communication crisis.[11]

Most people I talked to about the work asked me the same question, with varying degrees of sarcasm or candor: What is a bot? I still ask myself that simple question. The term "bot" encapsulates a lot of different sorts of automated online software programs. Rather than focus on bots writ large, however, I decided to focus on the use of social media bots—ones built to mimic real users—for political discussion on the internet. And eventually I became bold enough to argue that political bots are actually a new form of media that is changing the way online speech occurs and how information flows during crucial moments in public life.[12]

My initial research was on how governments and other powerful and well-resourced entities use these real-seeming political bot accounts to massively boost fringe political platforms, harass political opposition and minority voices, and trick reporters and the public into believing disinformation. Even a group of just twenty individuals can have a serious effect on political discussions on Twitter if they are constantly active, tweeting hundreds of comments a day. A group of 5,000 coordinated political bots can amplify such efforts, substituting sheer numbers for human nuance. Any such modestly large automated army can successfully redirect the flow of information toward the ends its deployers desire. I saw that the future of politics lay not in grassroots community organizing, but in astroturf organizing—falsely generated political organizing, with corporate or other powerful sponsors, that is intended to look like real community-based (grassroots) activism. Bots were my entry point into a world of digital deception where all manner of people make use of new technology to prioritize their version of the "truth."

Clues on the rise of computational propaganda were scattered throughout news articles, blogs, and social media posts that pointed to several different groups in several different countries. I began reaching out to those groups. At first, I had little luck getting bot makers and purveyors of disinformation to talk to me. I talked to many hackers, like our friend in Budapest, who told me that they had heard of deceptive contracting firms that worked for various political and private entities to build and launch political bots. I was told that these contractors also sold other unscrupulous services, such as launching bespoke troll attacks, doxing journalists (releasing their personal addresses online), and even staging fake offline protests.

I got a big break when I met the small but mighty community of coders creating Twitter, Tumblr, and Reddit bots for democratically positive or neutral purposes.[13] They taught me a lot about how their creations operated, and they gave me leads on where to look for people using bots for less scrupulous causes. They also showed me that bot builders don't necessarily have to have deep skills in coding. Both lone coders and state-sponsored propagandists, including Russian government personnel, used free online platforms such as IFTTT (If This Then That, a site that allows users to write small programs in plain English) to create a wide variety of software programs—including political bots—for use over social media.[14]

Members of the public bot community told me that although sites like Facebook and Twitter had policies about using automated programs, the rules were flexible. When necessary, these guidelines were fairly easy to get around through basic experimentation. If a platform had restrictions against messaging more than once a minute, for instance, you could just build accounts that messaged after every minute and one second. The people building political bots and launching disinformation campaigns knew this too. They were—and still are—one step ahead of the companies working to stop them.

I began to see that political bots are built in a "grayhat" area—that is, all sorts of people, both criminal and not, and even folks with next to no coding skills, can create political bots. They do so for a variety of reasons, and many don't stop to consider the ethical implications. Some of the people I've spoken to say that they are trying to do something good for society. A few, however, freely admit to being "blackhats"—people who exploit digital vulnerabilities for their own malicious purposes. Most of the groups I know to have built the worst bot- and human-driven political propaganda campaigns—the really horrible and abusive stuff—are not coders by trade. Some work for political consultancies or communication firms, or they are members of extremist political groups. Many are mercenaries who operate outside of the countries whose elections and citizens they try to manipulate.

Until the months after the 2016 US election, Facebook and Twitter were dismissive of what my colleagues and I had discovered. Maybe they were too busy, maybe they were understaffed, or maybe they willfully ignored our attempts to share our research. My research team and others around the globe tried to give tech companies papers that detailed the computational propaganda campaigns on their platforms, but nothing changed. In the rare instances that I found myself at a topical conference or workshop with people from Google, Facebook, or Twitter, I would discover that they were policy team members or lawyers. They routinely either attacked my team's research methods or dismissed the problem, or both.

Fast-forward three years to a hushed congressional auditorium. Sally Yates, former acting attorney general, and James R. Clapper Jr., former director of national intelligence, are testifying at a hearing before the Senate Intelligence Committee on Russian interference in the 2016 US presidential election. They sit on either side of an easel holding a massive poster and infographic. It reads "The Russian Toolbox in the 2016 Election."[15] The first line item is "propaganda, fake news, trolls, and bots." Around this time, no surprise, the major social media firms began offering me and my colleagues jobs. They began asking us to consult on this "new" problem. But by then the issue had reached such a magnitude and was at such a tipping point that I thought I was more likely to have a nervous breakdown than actually make a difference if I were working inside one of these companies to solve the problem.

The Human Element

In April 2019, President Donald Trump had a closed-door meeting at the White House with Twitter CEO Jack Dorsey. According to the *Washington Post,* Trump used the meeting to complain to Dorsey about recent significant drops in his follower account on the social media platform.[16] The Twitter cofounder attempted to gently explain to the president that he suddenly had many fewer followers because Twitter had deleted hundreds of thousands of bot and spam accounts. The *Post* pointed out that "previously, Trump joined a chorus of Republicans in claiming that Twitter secretly limits the reach of conservatives, a tactic known as 'shadow banning' that Twitter has vehemently denied." The newspaper went on to explain that "the president regularly has raised fears about changes in his follower count."

It has now become common for politicians to complain about social media platforms' supposed efforts to tamp down their online support networks and digital influence. The simple truth, however, is that many people—especially well-known politicians—benefit from bolstered follower numbers on sites like Twitter. Though new rules from digital platforms are beginning to change follower counts, it remains the case that many of the most active social media accounts are automated. It's also true that these accounts are overseen by people—people who are attempting to cheat the system—and that places limits on how much content any given account can generate over a given period of time.

I used to tell people, "When you do research on bots, a lot of bots want to talk to you." It would have been more accurate to say, "When you do research on bots, the people who build those bots want to talk to you." The first several papers I wrote about the use of political bots, whether in the United States, Turkey, or Mexico, resulted in Twitter messages and emails refuting my findings that one account or another was a bot. The people who had created the accounts would claim, with righteous indignation, that they were not bots, they were people. But even the quickest of looks at the profiles in question would alert the average user to the fact that something was off. Perhaps they had created hundreds of thousands, even millions, of messages. Much of the time the messages were focused on one or a few political topics, or they trolled the same people over and over again. Or else the accounts messaged ten times a minute, every minute, every hour, every day.

Yes, a person built the account, but a person also linked that account to software that automated its messaging capabilities. The social bots and the humans who built them were, and remain, intertwined.

Today politicians in various countries around the globe claim that researchers who find botnets or human-led groups furthering their divisive causes are conducting hit jobs on real people who support them. But when I began studying computational propaganda, few people were at all wise to the fact that some groups were using these tools and tactics. The world still saw social media through rose-tinted glasses. Many people believed in Google's slogan "don't be evil," and thought that Facebook was edgy for wanting to "move fast and break things." Some still do. To the extent that researchers paid any attention to "bad actors" online, they focused on people hacking technological systems like voting machines and government websites. My colleagues and I were more concerned with studying how people hack public opinion using online tools. There was no playbook for studying what we wanted to study, and there still really isn't.

Like nearly everyone else, I am still learning new things to this day about the political uses of the internet and technology. I hear about a new and exciting app or digital tool almost every week, but there is a lot that I still don't know. Indeed, there is a lot I will never know. A huge amount of the disinformation and misinformation never makes it to real users. Many of these propaganda campaigns are designed and launched in attempts to give the illusion of popularity to particular ideas, such as hashtagged (#) content on Twitter. Propagandists continue to launch bot armies across multiple social media sites in order to manufacture false trends that then get picked up by the media and the mainstream. Many of the strategies for disseminating junk news and political conspiracy online are hidden behind proprietary code, as are the algorithms that social media companies use to prioritize one story or post over another. These strategies are also obscured by layers of data that only an elite few—most of whom work at the social media companies—can understand.

You can learn a lot just by analyzing publicly available social media content. With a little training, it is easy to see that a great deal of social media traffic is of dubious origin. It's far from in-depth data science, but it can work. In fact, it can be as simple as discovering regular patterns in the timing and repetitiveness of the content of posts on a site like Twitter or YouTube. By keeping an eye out for markers, including abnormally fast rates of posting, you can identify social bot networks by their repetitive statements and links to articles in the posts and networks of shell accounts that were built to follow and repeat one another's content. You may even notice especially high usage of these automated tools and digital manipulation tactics during political events.

The Problem of Access

In the early days of our study, my team quickly came upon a research roadblock. Twitter only allowed researchers like us to download a very small, and potentially scrubbed, amount of content on a given topic, or during a given period, from its application programming interface (API). You could pay for better access to what Twitter called the Decahose Stream (10 percent of randomized content on a topic) or the Firehose Stream (100 percent of randomized content). However, this access cost a huge amount of money, and there were serious restrictions on how the data one received permission to gather could be used or shared. Other organizations, such as Facebook and YouTube, shared next to nothing with academics. They still severely restrict who can access data and how. Before 2016, when these companies did allow access, it was in a haphazard fashion.

Even when you do get access to quantitative data, it can only show you so much. Digits don't really allow you to explore emergent social or technological phenomena in any sociopolitical depth. For this reason, I have regularly collaborated with data scientists to combine social media data analyses with ethnography—the scientific description of customs and cultures. I still spend time with various people who were working in positions related to social media bots and the other novel technologies that were beginning to be used for political control. My goal is to figure out the right questions to ask, and the proper levers to pull, before my colleagues and I dive into broad-scale data analysis.

The world of computational propaganda and digital deception is complex and muddled. It is a world where a wide variety of groups use social media for political marketing in a somewhat disorganized fashion. During my early research, many digital consultants used the tired analogy of the Wild West to explain the lawless state of political communication online. It was a space, they said, where anything was allowed and no one was held accountable. Social media "experts" who went from campaign to campaign, election to election, laughed at me when I asked if spreading online disinformation about the opposition or using political bots was a tactic. "Of course," one US campaign staffer told me. "We throw everything against the wall and see what sticks."

He and his colleagues told me that, ever since sites like Facebook and Twitter were founded, they had worked on numerous campaigns in the United States that had used bots, disinformation, and political spam to influence conversations during elections. They used smaller messaging platforms like 4chan to create and test propaganda smearing the opposition. The story or meme would usually feature the candidate in question doing something illicit or ridiculous. One meme widely circulated in 2016, for instance, showed Hillary Clinton boxing with Jesus against a background of flames. It was later tied to propagandists working for the Russian government. Such memes or stories often suggested that the candidate was tied to a conspiracy—the kind of far-out claim that people love to share on social media. From platforms like 4chan this informational junk food flowed onto sub-Reddits—in 2016, for instance, /r/The_Donald, /r/conservative, and /r/altright were popular spaces for far-right disinformation and conspiracy. From these spaces, users who don't know who is behind a given ploy continue to pick up memes and spread them on Facebook, Twitter, YouTube, or Instagram, where they have an informational snowball effect. It's astroturf politics at its finest.

Propagandists have told me that they can achieve this cascade effect without using bots. They pull together users on sub-Reddits like /r/pikabu and get them to follow one another on Twitter. They use Twitter's private chat feature to initiate conversations with thousands of people in one chat window, and then they time "organic" hashtag bombs, which are designed to get conspiracy and disinformative content to trend on the site. These propagandists don't need automation; all they have to do is coordinate enough real users to trick the site's curation system. Then the general public shares the idea, article, or meme and it goes viral.

Political propaganda on social media uses a lot of the same tactics as email spam. The only difference is that companies like Facebook have made it a hundred times easier by selling access to the perfect clients. Someone selling Viagra is immediately given access to late-middle-aged men. The same goes for political marketing using social media. One person I interviewed, a fast-talking digital marketing expert, told me that political communications had quickly become a huge source of income for firms like Twitter and YouTube. "Elections happen every year," he said. "Of course, these companies are working every angle, with every party and candidate, in contests around the world—they're agnostic about who actually wins. They sell highly organized access to specific demographics and interest groups."

Among the groups I discovered that were working to manipulate public opinion using social media, several stood out. The strategies they claimed to have refined for reaching out to voters ranged from simply using paid advertising on a site like Facebook to engaging in downright deceit by using coordinated fake political campaigns to spread disinformation. I first encountered Cambridge Analytica, for instance, in the spring of 2016 in another dingy basement, this one in a New York City computer store. At an event that I came across, hilariously and surprisingly, on Meetup.com, a small group of top-level employees from the firm made a presentation on the "psychographic" method of campaign advertising that they were using for Ted Cruz's presidential campaign. They openly bragged to the mostly empty room that they had used Facebook and credit agencies to gather data on 230 million Americans. At the time, much of what Cambridge Analytica said about using social media to successfully politically target individuals using the information they had willingly shared over the same online websites seemed like bluster. They said that they could use massive swathes of this personal user data to construct both political and psychological profiles of American voters. The information from credit reports, they told those gathered, could be combined with the topics people posted about or groups they followed on Facebook. This data could then be amalgamated and refined in order to send out individualized and hyper-manipulative political ads. In the coming years, if serious changes are not made to social media, Cambridge Analytica's one-time boasts will likely be realized as real strategy.

What Happened: The Past

What do Australia, Azerbaijan, Bahrain, Colombia, Ecuador, Egypt, Iran, Mexico, Morocco, Russia, South Korea, Syria, Ukraine, the United States, the United Kingdom, and Venezuela all have in common? The answer: citizens in each of these countries experienced malicious political attacks and automated propaganda campaigns over social media prior to 2013. Politically oriented bots and human-run cyber-armies pushed propaganda, amplified particular points of view, and suppressed the ability of the opposition to organize over social media during pivotal political events. Online communication was heavily manipulated during elections won by the slimmest margins. Disinformation about political crises flowed at a staggering rate.

Nowhere was this more apparent than in North Africa and the Middle East. In an April 2011 piece for the *Guardian*, the free-expression activist

Jillian York wrote that "in Morocco, Syria, Bahrain and Iran, pro-revolution users of the site have found themselves locked in a battle of the hashtags as Twitter accounts with a pro-government message are quickly created to counter the prevailing narrative." Researchers and experts argued that actors associated with embattled regimes—whether hacking groups like the Syrian Electronic Army (SEA) or shady digital PR firms—were using social media to stifle online communication and organizing related to the uprisings.[17]

By 2013, the rest of the world had all but forgotten the democratic furor of the Arab Spring. The public spotlight had moved on to other things—Facebook's IPO, the London Olympics, and Obama's reelection. In Syria, though, what had begun as a democratic uprising had devolved into a roiling conflict over the future of the government. While the rest of the world watched Usain Bolt cruise into the record books, the people of Syria saw their country fall into a brutal revolution. The war was fought on multiple fronts, not only because it involved several actor groups with various geographic and sociopolitical allegiances, but also because it was waged both offline and online.

Sites like Facebook and YouTube became a battleground for a divided Syria. Armies of Twitter bots harassed anti-regime activists, suppressed online dissent by barraging opposition hashtags with spam, and amplified the positions of President Bashar al-Assad. The Syrian Electronic Army was at the forefront of these campaigns. The hacking group suppressed global awareness and altered local public opinion by simultaneously using a combination of social media, political bots, memes, targeted spam, and video. According to the FBI, the "Syrian hacker collective… hijacked the websites and social media platforms of prominent US media organizations."[18]

At the same time online political manipulation in Syria was rising, there was a broader international spike in the use of similar online tactics for coercion and control. Anti-PRI (Institutional Revolutionary Party) activists in Mexico were being demobilized over Facebook, Kurdish journalists in Turkey were being harassed by government supporters on Twitter, and US political candidates were being smeared with disinformation on YouTube.[19] By all accounts, from both people on the ground in these countries and current as well as former employees of large technology companies, the social media companies were doing little to nothing to prevent this misuse. In fact, many of the attacks that were being leveraged against the citizenry in these countries were being made through legitimate tools and services offered by the social media companies. Anyone could buy targeted advertisements on the sites, and early on the companies did very little to limit what advertisers could say or do. Anyone with the money could target users by race or ethnicity, for instance. Social media companies still hadn't changed their policies by the 2018 US midterms—two years after the debacle of Russian interference in the disastrous 2016 US "fake news" election.[20]

In fact, the line between legitimate political marketing by campaigns and digital propaganda was—and remains—almost nonexistent. In 2013 all manner of groups were using social media, online search, email, and other online media as experimental testing grounds for new strategies to capture and keep attention on the screen and—consequently—the political message. They were playing with the idea of individual- and group-specific political ad targeting. If they could hit people on issues that meant the most to them individually, then they could better manipulate their vote. These tactics are still being used. According to researchers Daniel Kreiss and Shannon McGregor:

> *Technology firms are motivated to work in the political space for marketing, advertising revenue, and relationship-building in the service of lobbying efforts. To facilitate this, these firms have developed organizational structures and staffing patterns that accord with the partisan nature of American politics. Furthermore, Facebook, Twitter,*

> *and Google go beyond promoting their services and facilitating digital advertising buys, actively shaping campaign communication through their close collaboration with political staffers. We show how representatives at these firms serve as quasi-digital consultants to campaigns, shaping digital strategy, content, and execution. Given this, we argue that political communication scholars need to consider social media firms as more active agents in political processes than previously appreciated in the literature.*[21]

Though most of the money spent on advertising by traditional electoral campaigns still goes to traditional media (TV, radio, print ads), for a variety of reasons a constantly growing amount is being directed toward the web. Campaign managers understand that more people use the internet, and tools like social media, than ever before. People also go online to learn about current events and engage in conversations about public life. For instance, according to a 2018 report from the Pew Research Center, over two-thirds of Americans get their news from social media platforms.[22] The big online platforms, from YouTube to Facebook, also offer campaigns the ability to target voters with a previously unthinkable degree of precision. These companies have more behavioral data—information on what people like, support, talk about, and hate as well as on where they are physically, what they look like, and which sociocultural groups they associate with—than any previous group of organizations in history. They can deploy this data, for a price, on behalf of campaigns hoping to reach highly specific subsets of a given population.

There is something else that campaign managers know. While traditional media sources are highly regulated, especially when it comes to political advertising—because democratic governments have had decades to devise ways to control what politicos can say using TV and radio and how they say it—the internet, by contrast, is anyone's game. The "Wild West" metaphor used by my sources is an apt one. "The world's media channels," according to the Organized Crime and Corruption Reporting Project (OCCRP), speaking of all forms of media, including those with a history of regulation, "are rife with propaganda, misinformation and simply wrong information."[23]

With just a little searching, the average citizen can find social media followers that can communicate in any number of languages, usually in the form of bots, for sale. In fact, a colleague and I once did a segment with NBC showing just how easy it was to buy social media bots.[24] Not only were these bot armies cheap to buy—thousands of follows, likes, or retweets often cost only $25 or even less—but they were also available for multiple social media platforms and in multiple languages. The web has a peculiar ability intrinsic to its communication configuration: virality. Things that gain popular traction online—even using bots to give the illusion of favor—can spread very quickly. In fact, in a study published in *Science*, MIT researchers found that false news and lies spread significantly more quickly online than truthful information.[25] Moreover, viral trends often get picked up by news agencies that then re-report them to the public. So even if someone doesn't use Facebook or Twitter, they often learn about what happens on those platforms through other channels. Digital propagandists understand these things and capitalize on them.

What Has Changed: The Present

Today social media companies, policymakers, academics, and civil society groups are scrambling to address the problems presented by computational propaganda. The mood is particularly frantic in government and at technology firms. Those in charge of regulating speech and protecting society and those who build the technology now being used to challenge democracy have been the most blindsided by the rise of disinformation. Policymakers in democratic countries have become so embroiled in issues like identity politics and the rise of nationalism, as well as the ever-increasing demands of political fund-raising, that they have failed to plan for, acknowledge, or deal with the growing revolutionary political power of the internet and social media.

Throughout the relatively short tenure of their innovations, technology makers in Silicon Valley, particularly at big firms like Facebook and Google, have suffered from technological solutionism and determinism. Their techno-utopian mind-set has prioritized new computational tools as cure-alls for society's ailments. This philosophy allows engineers to believe that advanced software and hardware are not only solutions to a problem but the mark of a sophisticated culture. The difficulties associated with this ethos—those at the heart of digital propaganda—are further exacerbated by a deep vein of libertarianism in the community, which has historically allowed tech leaders to disclaim responsibility for the issues that rear up at the intersection of the technologies they produce and the societies they inhabit.

In the face of computational propaganda, the social media companies' recent defensive motto of "we are not the arbiters of truth" begins to sound strangely childish. In essence, the firms are saying: "We built these things, but we are not responsible for the problems they cause." Google's "don't be evil" slogan seems at odds with this attitude, while Zuckerberg's "move fast and break things" seems in line with it, but these phrases are actually different sides of the same coin. On the one hand, companies have touted themselves as societal saviors, and on the other, they have viewed their work as inevitably transformative.

If technology firms have been too focused on tools as solutions, politicians have been overly embroiled in social and political issues absent any meaningful consideration of technology. Just as they have begun to attempt to respond to computational propaganda, it has become clear that politicians the world over weren't totally wrong to fixate on identity politics and nationalism as fragmenting issues for democracies and other polities. Nevertheless, in focusing most of their attention on these issues, they have failed to examine the impact on those issues of the slew of new informational problems posed by the rise of social media. Technologists have not been wrong to look to computational mechanisms for help in the face of these issues. But both social issues and technological ones were behind the rise of computational propaganda, and mechanisms from both domains will be needed to counter its effects.

It may be unsurprising that those who make computer-based tools would think of software and hardware as the be-all and end-all, or that politicos would see policymaking and reelection as their sole purview. These perspectives, however, reveal a deficit in the perspectives of our leaders, whether they are politicians or business executives. It's no longer the case that a legislator can succeed with no more than a law degree, or that a software engineer need only study computer science. We need transdisciplinary leadership and education—scientists who understand social problems and policymakers who understand technology. We need public interest technologists and technologically savvy politicians.

What Is Coming: The Future

By now almost everyone is familiar with the term "fake news." You cannot open a newspaper without reading about it. We still regularly hear about Russia's use of social media to manipulate voters during the 2016 US presidential election. That contest left Americans with little doubt as to the rise and repercussions of online disinformation.

But the problem is only getting worse. Far from having abated after Trump's win, computational propaganda has continued to spread at a rapid rate around the globe. What does the future of fake news—or more precisely, false news—look like? What technologies will be used to spread it? How will not just social media but also virtual reality, augmented reality, human-sounding automated speech programs, doctored videos (deepfakes), video games, and increasingly interactive online memes be used to drive political harassment and further polarize communities?

Signals of how technology will be used to drive new forms of computational propaganda abound. Ordinary people will be able to use new technologies to target those they disagree with, but it will still be powerful political actors—militaries and governments—that have the resources to engage in the most targeted and damaging forms of disinformation and harassment online. What is more, the manipulative usage of social media and new immersive tools will continue to result in offline violence.

In Myanmar, for instance, rumors about the Muslim Rohingya minority were started and fueled on Facebook by members of that country's military, leading to the murders of tens of thousands and the displacement of hundreds of thousands of people to date.[26] In India, disinformation campaigns on the mobile instant messenger application WhatsApp have led to murder, assaults on women, and attacks on journalists.[27] Throughout Europe, far-right parties have used Twitter to spread false online stories about refugee violence against women to propel isolationism and fear.[28] Less than twenty-four hours after the 2018 school shooting in Parkland, Florida, a conspiracy video claiming that certain outspoken survivors were "crisis actors"—trained performers portraying disaster victims—became the top trending clip on YouTube.[29] And propaganda about the 2020 US elections and other pivotal political contests is already being spread online.

But the greatest lesson of the past few years lies not in the tactics employed by online propagandists, but the speed with which they change. As I've said, research on computational propaganda was ignored by many tech companies until late 2016. It is impossible to say what difference it might have made to the outcomes of that year's elections if those companies had taken action, but surely their inaction is not a mistake we want them to continue repeating. That's why it's so urgent that we begin preparing now for the possibilities presented by emergent technology, like VR and deepfakes. As technologies like VR become tools for spreading defamation and disinformation, we will have a new set of problems on our hands. As one colleague put it to me, the body, unlike the eyes and ears, has no metric for fake.

As highlighted in story after story from my research, the problem is global, and so are the technologies. I've spent time discussing the rise of Chinese propaganda with technologists while in Japan and Singapore. I've had conversations about social media regulation with European politicians in the taxidermy-filled basement of a chateau in Slovakia. Bot builders and hackers have told me about gaming Twitter's trending algorithms from their homes in Brazil. One need look no further than the US special counsel's investigation into how Russia interfered in the 2016 US election using social media to get a sense of what transnational digital propaganda looked like in years past and guess at what it may look like in the future. The Kremlin has been the foremost innovator in the field of computational propaganda, but if China's Communist Party ever decides to replace its "50-Cent Army" of human propagandists with tools driven by AI and automation, the geopolitical impact could be greater than we ever expected.

We must all ask ourselves some serious questions about the future of computational propaganda. The solutions to this problem must come not just from governments and technology firms, but also from users and real people—all of us who have the biggest stake in making sure that future generations don't live in a world where reality and fiction, truth and lies, are indistinguishable. First, we need to think long and hard about what the future holds for digital propaganda and the emergent technologies that will be used to spread it. How might these same technologies be used to fight disinformation? What social solutions can be found? How much should we trust technology, specifically AI, to help us solve this massive problem, and how much should we depend on actual people to track and prevent the flow of disinformation and hate online? Which governments, organizations, political parties, and interest groups are emerging as the most dangerous and sophisticated actors in this new form of information warfare?

The Media Breakdown

In 2016 I spent a lot of time in New York City. It was the home base for both the Clinton and Trump campaigns, so I wanted to be where the action was. I had also taken on, simultaneous to my work at Oxford, a part-time research fellowship at Jigsaw, Google's human rights–oriented technology incubator located near the bustling Chelsea Market. I went there to do globally focused work on political bots and inorganic information operations—fake online attempts, both bot-driven and human-led, to boost particular content, especially during elections and big events. I'd travel to New York once every month or two and stay for several days. During each trip, I would schedule meetings and research interviews with experts based there, such as campaign staff, digital marketing consultants, journalists, and law enforcement officials.

On one trip I was lucky enough to be invited to talk with a high-level employee at a major US newspaper based in the city. (I am not providing names here because I promised confidentiality to this person and to all of my other interviewees during that time because of the sensitive nature of the political events taking place.) When I arrived at the offices, we went up to the cafeteria and, over coffee, began to chat about propaganda, bots, and the presidential race.

The newspaper had clearly worked hard for a number of years to incorporate social media into its reporting. It also used sites like Twitter, Facebook, and YouTube in efforts to more effectively connect the newsroom with readers, with varying degrees of success. We talked about the challenge of prioritizing information in the age of new media, especially given the rise of social media bot traffic and the heaps of bad reporting and rumormongering out there. We also discussed the challenge posed by social media algorithms, which prioritize certain types of news over others. These unknown bits of code were also beginning to curate certain information for certain users. Was that causing a rise in echo chambers? Were these moves enabling bots to artificially manufacture trends by making a bogus story look much more popular than it actually was? Beyond this, we talked about the threat to the foundations of the traditional news media of the success of social media platforms in both selling advertisements and getting users to consume news on-site.

This conversation, and several others like it with other veteran reporters and editors, crystallized something for me: not only has the rapid rise of sites like Facebook and YouTube been damaging to how we consume information, because of flaws in the design of these systems, but these sites have also usurped the role of traditional news media, because they present their media products as improved alternatives to old-school news outlets. Social media firms do this, both explicitly and implicitly, by simply reposting or creating "trending" material already produced by newspapers, magazines, and TV stations. The end result? Tech firms make advertising money off solid news and reporting content from other organizations.

Facebook and Twitter, unlike an organization like Yahoo!, do not produce their own news. But they certainly massage where on their sites other people's reporting appears, when it appears, and how it appears. They point to this method to underscore their argument that they aren't media companies (not "arbiters of truth," as I mentioned earlier) and to suggest that they are simply technology firms or service providers à la AT&T. But this disingenuous claim isn't true. Despite the fact that Facebook and Twitter don't employ journalists or write their own content, their algorithms and their employees certainly limit and control the kinds of news that over two billion people see and consume daily. They do arbitrate information, and by doing so, they arbitrate truth. Like the news media of old, social media companies set the agenda for what we get to see and how we see it. They frame our news in a particular way.

It's time for multibillion-dollar tech firms to partner with the news media in order to support professional, well-vetted reporting. We must not accept recent arguments made by some politicians that demonize the free press. As a scholar of political communication, I can tell you that these types of assaults are early warning signs of authoritarianism—of the rise of despots and monolithic regimes. The press is imperfect, but good reporting abides by strict guidelines and ethics and should be celebrated by anyone who supports freedom and our right to quality information. Google has recently launched a multimillion-dollar initiative to work with news organizations, and this is certainly a step in the right direction.[30] But there is still a lot more to be done, and other companies need to get on board.

Chapter Three
From Critical Thinking to Conspiracy Theory

At one point, the story was shared on Facebook over one hundred times a minute.[1] The all-caps headline, sensational and imminently clickable, read: "FBI AGENT SUSPECTED IN HILLARY EMAIL LEAKS FOUND DEAD IN APPARENT MURDER-SUICIDE." The news outlet, "the *Denver Guardian*," billed itself as Colorado's oldest news source. The story, written and released just days before the 2016 election, detailed a conspiracy implicating the Democratic presidential candidate and the broader party in a ghastly political cover-up.

But "the *Denver Guardian*" wasn't a real news site. And the article was totally false. The creator, Jestin Coler, was later found out to be a US citizen living in California. He had purposefully created the site and others like it in an effort to get ad clicks by spreading false news reports during the 2016 contest.[2] He was making money, between $10,000 and $30,000 a month, by capitalizing on the public's seemingly voracious appetite for junk news.

"The people wanted to hear this," he told NPR. "So all it took was to write that story. Everything about it was fictional: the town, the people, the sheriff, the FBI guy. And then...our social media guys kind of go out and do a little dropping it throughout Trump groups and Trump forums and boy it spread like wildfire."

As the one hundred shares a minute suggests, the story went viral. It became part of an internet-wide conspiracy. It wasn't spread, like other disinformation campaigns, by a foreign government working to undermine US democracy. It was spread by a fairly regular guy living in the California suburbs. But during the 2016 election, it was part of a larger digital offensive against the truth emanating from a disinformation ecosystem that included bogus Russian ads on Facebook, fake stories peddled by teens in Moldova, and political bots deployed by shadowy super PACs—allegedly independent "political action groups" in the US that can raise unlimited amounts of money from corporations and other entities (unions, wealthy individuals, etc.) in order to push their own agendas—over Twitter. Coler's website and arsenal of false news reports were capitalizing on people's desire to think deeply, tugging on the same cognitive mechanisms that spur critical thinking. But Coler and others like him were dressing up lies to look like news, leveraging social media and political garbage to spread conspiracy theories.

From Silicon Valley with Love

Although Special Counsel Robert Mueller's investigation into Russian interference in the 2016 US election helped to keep computational propaganda on the map, the computational propaganda used in that election actually came from a variety of domestic and international sources. Nevertheless, the investigation was a constant reminder of the digital disinformation and manipulation campaigns waged during the contest. It kept the attention of reporters—and by proxy, the public—focused on the broader threats to democracy of digital disinformation. Had this inquiry never been launched—or if it had been sunk early on by the Trump administration—the zeitgeist might have made it all too easy to let go of worries about things like false news and the serious shortcomings of social media companies. But the fear that Russian government employees and consultants might have tampered in a highly contentious American election—a reality eventually supported by the facts unearthed by the investigation—has thankfully kept our collective cultural sights on this serious and persistent problem.

Russian attempts to manipulate public opinion in 2016, however, are only the tip of the computational propaganda iceberg. Although transnational political attacks during elections, illegal under both US and international law, are especially worrying, it is also true that individual people spread disinformation and politically harass other individuals, using social media and other technology. It also happens between all sorts of groups, from corporations to civil society entities. Individuals spread disinformation and harassment both across international borders and within the boundaries of one country. Such uses of media technology to manipulate people into seeing the world in a particular way have a long and complicated history, one tied to the longer evolution of the news media.

Before 2016, social media platforms had their own short history of being used as tools for political control. You know by now that a profusion of politically oriented offensives were launched over major online platforms going as far back as the 2010 Brown–Coakley special election in Massachusetts. Some people argue that such machinations go back even earlier, to around when Twitter was founded.[3] Perhaps most of us just weren't looking out for disinformation at that time. Researchers have established, however, that hackers and other groups were using chat applications, including precursors to modern social media such as internet relay chat (IRC), to plan offensives and engage in political dissent.[4]

Some might argue that social media firms and other technology firms can be forgiven for buying into the hype about their products being free and open tools for spreading and supporting democracy and unfettered communication. The relatively short history of Silicon Valley is defined by stories of techno-utopianism and cyber-libertarianism, and many of today's social media tech companies capitalized on this ethos.[5] But it is important to not forget that Twitter and Facebook were being used to interfere in elections a decade ago.

Why weren't social media companies paying attention back then? Why weren't policymakers amending the law to protect citizens from artificially amplified online barrages of politically motived hate speech and false information? Why weren't US politicians, whose jurisdictions include many of the biggest social media firms' home offices, regulating the political advertising space on social media? Why aren't they doing so now?

Journalists have been reporting on the political misuse of social media since well before 2010, so the information was certainly out there. Technology companies and governments failed tremendously in not dealing with the problem of computational propaganda when it first reared its head. Undoubtedly it would have been a costly problem to address then, but it is certainly a multibillion-dollar catastrophe now.[6] And money is no excuse for not addressing an issue that now threatens the very fabric of democracy. So what happened?

From Online Utopia to Digital Dystopia

The offices of the Electronic Frontier Foundation (EFF) are in a multistory house in the middle of San Francisco. The organization describes itself as a donor-supported nonprofit entity that "champions user privacy, free expression, and innovation through impact litigation, policy analysis, grassroots activism, and technology development." Founded in 1990, the EFF is well-known in the tech and intellectual property communities as a proponent of the free and open internet. "Even in the fledgling days of the Internet," its website's "About" section reads, "EFF understood that protecting access to developing technology was central to advancing freedom for all." The group has done laudable work in supporting open-source software—programs with open, editable, and nonproprietary code. It has supported the rights of individuals and small companies and continues to argue against the privatization of the digital sphere.

But the EFF, like other organizations concerned with the protection of digital civil liberties, has long faced a challenge that has become more acute since the rise of web 2.0: massive companies, like Google and Facebook, now dominate a huge percentage of internet traffic worldwide. Not only do these organizations own the products that allow people to connect with one another, but it is their proprietary code that determines who gets heard and how. Their algorithms decide what goes viral and what news people read. These firms, in many ways, control the right to free speech online. While they may say that they promote a fair and humane internet communication space, their actions suggest otherwise. These companies have revolutionized the information ecosystem and created new and complex questions about free speech and other democratic norms. These changes have come to a head with the rise of computational propaganda.

It was just about thirty years ago—right around the time the EFF was forming—that the internet went from being a government communication tool to a hybrid public–private commodity. Our lives have changed drastically since then. The way we share information has gone from one-to-one tools, like the telephone, and one-to-many services, like the evening news, to a vast many-to-many multimedia system in which everyone is a newsmaker or pundit. Media experts' fear in the late 1990s and early 2000s about large-scale consolidation of traditional media organizations has changed to today's concern about the ability of regular people to influence opinion through the megaphone of social media.

The internet has facilitated a shift in who can produce "news" and how, but the news media landscape remains an absolute mess. Large and strategically unwieldy conglomerates still maintain control over broad swathes of broadcast, radio, and print media. Emergent digital news organizations like *BuzzFeed*, *Huffington Post*, and *Vice*, still struggling to come up with a viable business model to compete in a frenetic news environment, have recently had to seriously cut back on staff.[7] They and their competitors are not quite separated from the previously dominant media business model that necessitated high-rent, high-overhead New York City offices. They have not been able to effectively realize a nonhierarchical, remote business model—if that is in fact a solution. Nor have journalists in general used their profession's historical experience of grappling with technological changes to successfully ride the wave of the internet, as some suggested they might do.[8]

MOCK NEWS FLASH

We interrupt this virtual reality experience with breaking news from Digi-national Public Reporters (DPR): the Chicago Sun-Times, the San Jose Mercury News, and the Seattle Times have all been acquired by DPR in a bid to expand lightning-fast, hyperlocal, objective reporting to the digital sphere. Users on the reporting hub Twitter, the formerly public social media company now owned by the New York Times Company, are reacting favorably to the news. The "certified bot-free"™ trending algorithm on the site shows #DPRforDemocracy as the number-one shared hashtag on the platform. It's a brave new virtual era for reporting out there, folks, and don't forget: "Digitocracy dies in darkness."

Even if the global news-making model could be changed to something nimbler, it would still be extremely hard to effectively report on news in Pakistan or on Wall Street without being there. Some entities wagered that the internet would make this kind of remote work possible, but firsthand human experience is often still the best evidence for journalists, researchers, and the general public. News companies have made attempts to implement "satellite" organizational models that prioritize remote employees. In fact, these moves often coincide with efforts to shift staff reporters to shorter and shorter contracts, resulting in a corresponding loss in the level of institutional support and reporting resources provided to them. Satellite models also fundamentally alter—and not for the better—organizational provision of health-care benefits and steady pay. People are well aware of what these changes signal: news organizations are dropping like flies. What many people don't know, however, is that social media companies bear a healthy portion of the blame for their demise.

Facebook and YouTube came onto the news scene in the early 2000s shooting from the hip. They had made exciting products that could give people large-scale access to data and connect them to broad social networks. But they didn't plan on their systems becoming the world's default means of accessing information—of accessing the news. Ten years later, with the abrupt arrival of computational propaganda and junk news on the international scene, these firms realized that they had opened Pandora's box. They now had to grapple with the simple fact that their opaque algorithms were curating and consolidating information on everything from weather to breaking news. With other online information monopolies, they had effectively revolutionized and monopolized advertising-based models based on surveillance capitalism and driven by gargantuan troves of data on user behavior.[9]

Old-school news revenue models have been left in the dust. Companies like eBay and Craigslist twisted the knife further by eliminating most people's use of the newspaper revenue stream that was classified ads.[10] Simultaneous to all of the changes in the media landscape—and in many ways because of them—people have gained the ability to talk with almost anyone, instantaneously, across the world. With Twitter, everyone is a journalist, or so the saying goes. Although the internet has become ambient background noise in our everyday lives, it continues to transform how we think, feel, speak. It has challenged how we navigate life—literally down to our ability to comprehend directions in the age of GPS and Google Maps.[11]

The internet threatens to change our mental models—re-forming how we see and understand the world. It damages our perceptions of truth, trust, and fact. No longer do captive audiences get their information on local and global affairs from respected nightly news anchors or other journalists who make concerted efforts to be objective. Now people pick and choose their news product—and in some cases, their disinformation—from any number of millions of websites and social networks. The broadcast news era was far from perfect, but what we have now is markedly more complicated. It's also even more consolidated in terms of ownership and market share. The old media monopolies gave way to new media plutocracies that are even less transparent in governance and function.

The rise of our hyperconnected society, and the vast technical infrastructure underpinning it, has caused revolutionary changes to economic, political, and cultural systems around the world. Social media, and the internet more generally, were originally envisioned as utopian tools for spreading free speech and strengthening civic participation, but they were quickly co-opted by states and other powerful political entities seeking control of them.[12] It's true that everyday people can still use these sites to report on breaking news, to catch stories even before trained journalists find them, or to debate politics with a range of different people. But it's also true that politicos and the people have battled over the "wealth" of these online networks since before the web went public.[13] Indeed, the web has become as much a tool used to control people as a means to connect and empower them.

You Are What You Read

Seventy thousand students sit in one of many massive, virtual auditoriums. A professor, appearing as a purple-haired avatar of Albert Einstein, lectures the digital class on the fallacy of vaccination. He tells the class about the well-known government conspiracy to selectively infect the public with disease as part of a broad-scale eugenics practice. He shows charts and graphs detailing what he says is a correlation between the advent of vaccines and a massive growth in publicly acknowledged instances of autism. He pontificates on research that proves the flu vaccine was created to give people the flu, not to cure them.

This is, of course, a completely fake scenario. But it is not far from today's truth. Organizations like PragerU now produce high-quality, pseudo-academic videos and other digital material to combat "leftist indoctrination" on college campuses.[14] The content, viewed online over a billion times by people (and, most likely, bots) as of 2019, is both highly partisan and largely disinformative. Some of it is presented by well-known establishment journalists, albeit those with a far-right bent, like Charles Krauthammer and Steve Forbes. Most of the content in the videos of the unaccredited "university" flies in the face of science. Social media makes it possible for unreality like this to become canon in certain groups both online and off. The entities that create and spread this stuff know that they need social media to thrive and know how these digital spaces work. According to *BuzzFeed*:

> *PragerU spends a lot of time and money figuring out how to push people to its content. It spends more than 40% of its $10 million annual budget on marketing. That includes targeted campaigns on Facebook and Instagram, preroll ads on YouTube, and a 1,250-strong high school and college student volunteer team called PragerFORCE, who flog Prager content on their social feeds. PragerFORCE members aren't paid, but they are rewarded with shares from the main PragerU Facebook account, which has nearly 3 million subscribers.*[15]

It is largely changes in how we get information that have allowed the rise of digital disinformation and political manipulation, whether the content is spread by PragerU or other subjective outlets. Just as some democratic movements were gaining steam through the use of online technologies, the use of the same tools was stymying others. Still other movements, many nondemocratic, have come into being because of the networked power of social media.

While students at PragerU or in the fantasy virtual auditoriums tune in remotely, those using social media to engage in on-the-ground protest in Ukraine or the Middle East do not. The stakes are much, much higher for these users, who put their lives at risk in fighting back against autocrats and authoritarian regimes, whether they do so online or offline. State-sponsored political attacks against activists, which many have attributed to government groups, even take their tactics offline. People who speak out over social media in these places are threatened or even tracked to their own homes and murdered.[16]

Activists the world over need social media to communicate and organize, but they face more challenges than ever in their attempts to do so, owing to the rise of computational propaganda. The PragerU-type organizations of the world contribute to the flood of poor-quality and deliberately misleading content to people who otherwise have little access to higher education. These organizations are not only bending reality for people living in the crosshairs of despotic regimes but manipulating already vulnerable populations the world over.

My own research into how social groups and "issue publics"—those groups that engage in politics because of their interest in a particular topic, such as gun control or abortion—reveals that these groups are being targeted with computational propaganda now more than ever.[17] The biggest targets of online political campaigns are often those who are most defenseless or alienated—those who have no voice in mainstream politics but whose participation is integral to the functioning of any true democracy. Minority groups are particular targets for digital disinformation through social media–borne attempts at voter disenfranchisement and silencing.

Regular people are frequently conscripted as the foot soldiers for computational propaganda campaigns and often unknowingly drawn into operating on behalf of governments, militaries, and well-resourced groups. Citizens' willingness to believe in conspiracy theories more than the results of investigative journalism or scientific inquiry makes them susceptible to buying into campaigns that are against their own interests and the broader interests of democracy. If a democracy is governance by the people, then we all bear some blame for the current predicament. But we also have the power to change the current system. To do so we have to learn to distinguish critical thinking from conspiracy theory. We must demand access to high-quality news and information and figure out a way to get rid of all the junk content and noise.

From Critical Thinking to Conspiracy Theory

Bullshit caught the attention of the University of Washington campus, and the national news media, near the end of my doctoral career. In early 2017, Professors Jevin West and Carl Bergstrom offered the undergraduate course "How to Call Bullshit on Big Data." The online syllabus bluntly synopsized the class: "Our world is saturated with bullshit. Learn to detect and defuse it."[18] This was the year when the *Collins English Dictionary*, *Newsweek*, and the American Dialect Society all named "fake news" the term of the year.[19]

In the face of this, West and Bergstrom offered a course that made students question the provenance, prevalence, and purpose of information—especially digital information. In Seattle, the home of Microsoft and Amazon, students were set to learn about how lies and falsehoods could be dressed up using data and technological jargon and spread via social media.

We know now that hybrid data science and political communication companies like Cambridge Analytica played a significant role in making the public aware of "fake news." These companies' claims to have conducted psychographic marketing, dressed up in their marketing materials with pseudo-scientific language and allusions to big data analyses, were mostly bullshit too. Not only did Cambridge Analytica and myriad other firms like them rely on malicious propaganda tactics in their attempts to drive votes to the candidates they represented, but they used the language of data science and the smoke and mirrors of social media algorithms to disinform the global public. Some might argue that most people in democracies—especially the United States—are inured to ever-present negative campaign ads, but no one likes to be lied to or outright manipulated.

The course at UW was offered just months after Trump's win and in the midst of the burgeoning scandal around Cambridge Analytica, Facebook, and Russia. In the preceding years, "big data" and "algorithm" had become all but dogmatic terms in much of the tech sector and some parts of the academy.[20] This class was one small step toward setting the record straight. The professors offered a course that they hoped would dispel some of the mysticism around data science while also strengthening students' critical thinking. "While data can be used to tell remarkably deep and memorable stories," Bergstrom told *The New Yorker*, "its apparent sophistication and precision can effectively disguise a great deal of bullshit."[21] In this statement, the professor and data science expert sums up the current techno-dilettante-led assault on reality. Groups like Cambridge Analytica have attempted to equate the profusion of data that makes up the internet with the same data central to scientific knowledge. They have done so, however, without consideration of the other crucial elements of science—empiricism, verifiability, and replicability among them. They have used "data," broadly speaking, to give bullshit the illusion of credibility.

Many digital political consulting firms and a slew of social media marketing companies manipulate communication—not technology, data, or code—in order to peddle conspiracy theories rather than critical thinking. They use new technology, whether social media or VR, only as a means to getting to the largest portion of voters.

I spoke to one such social media consultant who told me that the absence of truth is as important in winning elections as the actual truth. When people are confused or angry, he said, they either don't vote or vote for the person who speaks to their anger. This same guy claimed to be agnostic when it came to questions of ethics or truth. When I challenged him on the morality of lying to voters to win a campaign, he laughed. He argued that it was impossible to really distinguish between good and bad, true and false. "That's politics," he equivocated.

But the texts at the heart of democracy—from Plato's *Republic* to Dewey's *The Public and Its Problems*—do attempt to discern right and wrong, especially when it comes to politics, communication, and the rule of law. There are inarguably many different perspectives on morality, but when it comes to what we can know here and now, what we can observe in today's world, the tenets of science give us a compass for distinguishing between true and false. Science can seem as difficult to translate in today's social media world, however, as it was in Dewey's analog world. Moreover, most of the institutions to which we once looked both for knowledge and as a moral touchstone are on the decline in the United States.

Someone inevitably asks me a question about the credibility of information every time I give a talk on computational propaganda. With the rise of public consciousness around "fake news," this isn't surprising. Trust in most US institutions is at an all-time low. According to Gallup, only 38 percent of Americans polled in 2018 had confidence in religious institutions, compared with 65 percent in 1973, when Gallup first polled on this topic.[22] In 2018, 11 percent of people had confidence in Congress, compared with 43 percent in 1973; 36 percent had confidence in the medical system, compared with 80 percent in 1973; and 30 percent had confidence in banks, compared with 60 percent in 1973. You get the picture. Gallup's numbers on confidence in US institutions reveal that, with few exceptions, trust has consistently declined over the last fifty years or so.

Despite the recent fervor about the modern "war on science," trust in the scientific community has remained stable since the 1970s. But as the Pew Research Center points out, this stability doesn't mean that most people actually trust science. In fact, when Pew last checked, only four in ten people in the United States said that they had "a great deal of confidence" in the scientific community. Moreover, other Pew studies have:

revealed that public trust in scientists in matters connected with childhood vaccines, climate change, and genetically modified (GM) foods is more varied. Overall, many people hold skeptical views of climate scientists and GM food scientists; a larger share express trust in medical scientists, but there, too, many express what survey analysts call a "soft" positive rather than a strongly positive view.[23]

What accounts for the decline in US confidence or trust in institutions? What about the corresponding decline in places like South Africa, Italy, and Brazil? Are people just becoming more skeptical? Are the institutions getting worse and worse at getting the job done? And what are the general implications for how we view truth? One poll of British people found that more than 50 percent worry about being exposed to fake news and 64 percent cannot tell the difference between real and false news.[24] Even more concerning, a different survey found that around 35 percent of Brits say they try to avoid the news entirely.[25]

Undoubtedly, many things are to blame. In an interview with the *Atlantic*, the CEO of the global communication and marketing firm Edelman, Richard Edelman, said that "the root cause of this fall [in trust] is a lack of objective facts and rational discourse."[26] He made this statement in 2018, just after his firm put out a comprehensive report that found only one-third of Americans trusted the US government, down nearly 15 percent from the year before. But as the *Atlantic* article points out, the public's trust is actually rising in another crucial part of the world. In China, a staggering 84 percent of people said that they trusted their government at "levels the United States hasn't seen since the early Johnson administration."

You may be thinking that, since China is an authoritarian regime and the United States is democratic, people *say* that they support the People's Party but might be too afraid to say they do not. This hypothesis undoubtedly bears some weight. But consider this: Freedom House—a global standard for ranking state freedom—recently downgraded political rights in the United States from a 1 (the best score) to a 2 (still mostly free, but not completely).[27] Freedom House is a respected not-for-profit organization founded by Eleanor Roosevelt and Wendell Wilkie that focuses on tracking global governance trends. It produces an annual report ranking state regimes and governments with aggregate scores on sub-categories including political rights and civil liberties. Prior to 2016, these reports used a points system from 1 (most democratic) to 7 (most authoritarian), but now uses a system from 1 to 100. Freedom House explains that "the United States' political rights rating declined from 1 to 2 due to growing evidence of Russian interference in the 2016 elections, violations of basic ethical standards by the new administration, and a reduction in government transparency." Others, including fascism experts Ruth Ben-Giat of New York University, Jason Stanley of Yale University, and the Southern Poverty Law Center's David Neiwert, have gone further, arguing that authoritarianism is on the rise in the United States.[28] The UK, meanwhile, still maintains "one" scores across Freedom House's categories of "freedom rating," "political rights," and "civil liberties" but has slowly declined in "overall aggregate score" (out of 100 points total) from 2017 (95), to 2018 (94), to 2019 (93).[29]

So governance models may not be the primary cause of declining or increasing government trust levels. This said, we still don't have a good argument as to why trust in government is falling in the United States and elsewhere while it rises in China.

Social Media Is the Message

As someone focused on the study of politics, the internet, and technology, I look to the media for the answer. The way we consume media throughout most of the world has changed a great deal since 1973. Back then people relied upon a few major TV networks, radio, their local paper, and possibly a big national newspaper to tell them about current events and the world. The internet has broken this system. Suddenly everyone can create content, and anyone can blog about the news. But in China and other authoritarian countries, the internet has been firmly controlled from the outset. Digital politics experts Shanthi Khalathil and Taylor Boas have detailed clear proof that numerous regimes have worked to diminish the web's information-sharing power by restricting access and closely monitoring in-country use. They suggest that "[some] uses of the Internet reinforce authoritarian rule, and many authoritarian regimes are proactively promoting the development of an Internet that serves state-defined interests rather than challenging them."[30]

So has open access to information via the internet caused a decline in institutional trust in the United States and elsewhere? Is the steady decrease in trust in some democracies due to people throwing off the shackles of big media and becoming more savvy about the efforts to control the previous—and still powerful—one-to-many media system? Did the internet truly herald a new age for democracy? Are we just more woke?

Yes and no. According to analyses from Pew, Gallup, and a panoply of other research groups, we are in the midst of working through a great deal of confusion about what is real and what isn't, who can be trusted and who cannot. The internet—along with big data, algorithms, and social media—has led to a great questioning of reality and trustworthiness. Many argued that it would be the savior of information equality, but they forgot to consider how an unfettered internet could be used for control. The web, just like any other technology or tool, is a vehicle for people trying to share information and ideas. It is media and it can be manipulated. No or little regulation of this novel media system in the United States—especially when it comes to the political use of social media—may prove as challenging for democracy as tight control elsewhere has proven to be.

The internet is obscured by unthinkably dense and complicated layers of code. Understanding this complex technological system of software and hardware requires not only proficiency in a wide variety of coding languages but also the ability to grapple with a convoluted system of governance. This is why those with the power and resources to decipher the system have most of the control. Beyond this, life online and the pursuit of digital information of any kind are often obscured by anonymity and amplified by automation. Of course, these same features of the internet and other technology have benefits as well as problems. Democratic activists can use anonymity on social media to evade strict regimes. Journalists can use bots to radiate information on a breaking story. But increasingly such benefits are at odds with the drawbacks. These tools have been used to undermine democracy and to make it impossible for some people to distinguish critical thinking from conspiracy theory.

In my research I often ask people how they decide what online news or social media content is real and what they think is fake. They mention everything from scientific analyses to their own political views and cite the accessibility of a wide variety of sources through new media tools, but they also concede that they tend to routinely access the same sites rather than look for new ones. This leads me to wonder about how people interrogate information using novel, far-ranging information technology. Perhaps critical thinking and conspiracy theory exist on the same continuum? They are very different, when all is said and done, but it is also true that they are related modes of thinking.

It is easy to get critical thinking and conspiracy theory confused. Both are geared toward "digging deeper" into an idea, argument, or occurrence, and both interrogate what seems—on the surface—real or true. During my public talks on computational propaganda, I regularly get questions about how I distinguish fact from fiction: How, people ask, do I distinguish between a real news article and a false news article? Aren't real news articles often subjective or sensational, the crowd asks? As someone trained in the social sciences, I usually point toward the need for evidence and data, interviews with firsthand witnesses, and an effort to be objective. But, the keen questioners will inevitably ask, can't interviewees be biased? Isn't all research always and already biased by the values of the researcher or journalist? This is where things get messy—and where people may leave the path of critical thinking to follow the path of conspiracy theory.

Take "QAnon," a conspiracy theory mostly spread online among elements of the far right in the United States. This theory revolves around the idea of the "deep state," a hidden intragovernmental force working against President Donald Trump and broader conservative values. Q, a member of the anonymous message board platform 4chan (hence the name "QAnon," the Q being a reference to top-secret "Q" clearance), started the conspiracy in the fall of 2017 with a post on that platform suggesting that they had classified evidence of forces working against Trump and the US government. Q worked to push vestiges of the "pizza-gate" conspiracy, the root of which was a claim that numerous liberal US politicians were engaged in a child sex ring being run out of the basement of a pizza joint. QAnon adherents attempted to tie "pizza-gate" to a larger deep state—a shadow institution led by Barack Obama, Hillary Clinton, George Soros, and others. Calling their movement "the Great Awaking," they argued that members of this "deep state" group were guided by illicit and far-left values and absorbed in efforts to hide their own lies. The problem with Q's claims is that hard evidence for them has never surfaced. In fact, strong evidence to the contrary has been released at every juncture of the dissemination of the QAnon conspiracy.

While it's important to note that not all conspiracies are fake, it is crucial to point out that the means and ends that characterize conspiracy thinking are flawed. This way of thinking tends to be driven by the idea that a powerful force or group is working to conceal some sort of secret related to control or coercion. People who think this way tend to want this to be the case—they are always seeking to unmask some hidden, structurally scaffolded secret. No amount of evidence—especially contradictory evidence—can negate the arguments of conspiracy theorists. The simplest answer is never the answer for conspiracy theorists, who will dismiss contrary evidence out of hand or incorporate it into their conspiracy—for instance, by suggesting that it indicates an elaborate ruse.

Conspiracy theory travels effectively, and often quickly, over social media. Anonymity allows people to express ideas that they wouldn't give voice to publicly. Anonymity gives Q the illusion of veiled intrigue, of a faceless patriot working to save democracy from one form of monster or another. Automated social media bots are deployed alongside coordinated groups of humans to seed and amplify conspiracy theories like QAnon around the internet, from 4chan to Reddit to Twitter and beyond. Social media companies have traditionally taken a hands-off approach to monitoring speech, whether in the form of conspiracy or legitimate political discussion, and done little or nothing to prevent the inorganic flow of such messaging. Pretty soon, regular people pick up the idea, the media begin reporting on it, and a full-blown conspiracy is born.

Conspiracy theory has arguably existed since the beginning of human civilization—that is, since power and rumor first began to affect human discourse. Today, however, social media has created more spreadable, more potent, and farther-reaching conspiracies than those spread by older forms of media. These effects travel offline. Around the globe, the dissemination of social media conspiracy has been followed by violence and death. While it's important for people to be able to discuss ideas freely, it is not all right—or legal in most democracies—for that discussion to incite violence, spread hate, or perpetuate slander. This is why social media companies must work to stop the flow of such information. A conspiracy theory that cites a bogus secret to argue for a violent response to a certain religion or race, for instance, has no more place on social media than it does on television. More than that, the people who use such flawed logic with the intent to generate violence should be kicked off social media platforms and prosecuted by authorities.

What About Policy?

We are deeply lacking in laws that would protect the flow of news and govern political communication online, whether that communication is aboveboard or hidden and manipulative. In the United States, the Federal

Election Commission made a sweeping decision in 2006 to all but ignore political campaigning online.[31] The FEC wrote that only online political ads fell under campaign finance law, but the agency has done a horrendous job of monitoring even that particular space. The government and social media firms have mostly relied—and still rely—on Section 230 of the Communications Decency Act of 1995.[32] This widely misinterpreted policy was intended to divest social media firms of responsibility for the speech that users engaged in using their platforms. If a neo-Nazi wrote an anti-Semitic diatribe on Facebook, for instance, the company wouldn't be culpable. Under Section 230, the US government gave internet-oriented corporations the right to censor harmful communication on their sites. But, crucially, it also passed off to the companies much of the onus of making tough decisions about free speech, without holding them responsible if they made an error in judgment.

Although they sometimes took the relatively clearer route of deleting content associated with violent extremism, the social media companies moderated disinformation and political harassment content in a very piecemeal fashion and took Section 230 as a license to do little or nothing about problematic political content. Executives at the companies pointed to it, confusingly, as evidence for their perennial claim that they were "not the arbiters of truth." In fact, the statute gave them license to arbitrate content on their platforms, but in the cyber-libertarian ethos that has long pervaded Silicon Valley, they chose not to act. They scaled sites blindingly fast without much consideration for ethical design or, God forbid, measures to prevent political misuses of their tools. There was little to no system for dealing with digital propaganda at any of the major social media firms prior to 2016. Even the system that exists today is ad hoc. In interviews with me, people at Facebook have described their company as a plane being flown while it is only half built.

The US government, particularly the Federal Election Commission (FEC), the Federal Communications Commission (FCC), and the

Federal Trade Commission (FTC), has failed to consider the possibility of social media being used as a tool to challenge fundamental democratic ideas. These regulatory bodies have ignored the possibility of these tools undermining not only cogent civic discussion but also parts of the voting process. The FCC and FEC seemed to have barely wrapped their heads around the concept of email just as social media burst onto the scene. The slow-moving leviathan of government failed miserably to regulate these tools, which quickly became the media of choice for people the world over, and their primary means of gathering news. Both the US government and Silicon Valley, to put it kindly, prioritized innovation over ethical rigor and caution. Put more bluntly, they prioritized economic and user growth over democracy. The mythical notion of "scale," constantly discussed by venture capitalists and start-up employees in Menlo Park, Palo Alto, and Mountain View, won out over human rights.

Internationally, other regulators have made more concerted and educated attempts to deal with the problem of digital propaganda. The European Union (EU) and some of the member states, including Germany, have led the charge, albeit in a rather heavy-handed fashion. In early 2018, German policymakers implemented a law that institutes heavy fines—up to 50 million euros, depending on the offense—on internet-based companies that fail to remove hate speech from their platforms. In 2019, German regulators passed laws that have crippled Facebook's ad business in that country.[33] In the spring of 2018, the EU rolled out General Data Protection Regulation (GDPR), a consumer data protection law that requires companies like Facebook and Twitter to be more open about what data they have on users and how they make use of it. This law has implications for digital political communication—from individualized political harassment and doxing to political ad sales on social media—but, again, the legislators who penned it often overlooked the feasibility of technical implementation in favor of surface-level reform. The EU correctly isolated and legislated on problems surrounding user data privacy. But, in reality, their efforts to force online entities to alert users about particular data gathering practices resulted in constant and heavy-handed notifications to all users rather than serious changes in how or whether data was actually gathered.

Current policy solutions for new problems associated with the age of data that have been enacted in the EU, Brazil, and other countries—such as GDPR and the Brazilian Internet Bill of Rights ("Marco Civil")—are moves in the right direction, but some experts have criticized them as too broad, possibly unenforceable, and hewing too close to censorship.[34] In other estimations, governments are applying Band-Aids to societal ailments that amount to internal bleeding. These allegedly simple responses are said to reflect no true understanding of how problems developed or what kind of remedy is needed. In the words of Rand Corporation researchers, policy efforts to fight disinformation use the "squirt gun of truth" to fight the "firehose of falsehood."[35]

Had they looked to history, politicians and technologists would have known that with great media innovation comes great change. The creation of the printing press in the Middle Ages, for instance, was followed by nearly two hundred years of propaganda operations by the Catholic and Protestant Churches. Social media have allowed for this type of knowledge-based conflict, but with computational enhancement, pervasive anonymity, and the automation of communication. False news has a long history—but those in power and those at the forefront of computer science failed to heed it. Today it seems that pundits and politicians alike are capitalizing on conversations about "fake" news—using the fear and confusion surrounding the internet's role in causing polarization and distrust to further their own agendas. Consequently, social media companies are continuing to scramble to convince regulators and the public that they are not the arbiters of truth—that they are technology companies, not media companies. They are losing this battle, but governments don't seem to be picking up the slack.

Media-Oriented Solutions

Because many of the social and political problems caused by social media are ongoing, a great deal of the discussion about media-oriented and policy-oriented solutions to the problems of computational propaganda is backward-facing. As Twitter gets a handle on its political bot problem, junk news is already being spread by the new hybrid automated-human forms that are on the rise there. In these new cases, many Twitter users are combining automatically functioning accounts with close human curation rather than using simple bots clearly trackable by extant tools. This move allows them to still post at inordinate rates while avoiding account deletion. As Facebook works to rein in the scope and scale of political advertising on its platform, the people who design information operations are moving onto group pages and other areas where they can interact with users directly. As YouTube stops linking users to videos containing conspiracy theories, deepfakes proliferate.

And these are just the attempts of "legacy" social media to combat computational propaganda. How are disinformation and unreality being spread on new social media sites or on encrypted chat applications? How are new tools being used to game the truth? Following Facebook's $3 billion acquisition of Oculus in 2014 and a slew of other big moves in the space around that time, hype around VR is fading.[36] But the technology continues to become more available, and in a variety of different forms. As tech investor Andy Kangpan points out, "The virtual reality market has continued its slow march to mass adoption, and there are tangible indicators that suggest we could be nearing an inflection point."[37]

Might VR be the next space heralded as a forefront of innovation and then quickly realized as one for control of public opinion? How will we get our news on VR social media platforms? Will increasingly sophisticated machine learning chatbots be built into these emerging communication technologies? What happens when chatbots begin not just to write like us but to look and sound human as well? It's easy to forget that 2016 was not just the year of Cambridge Analytica and Russian manipulation; it was also, according to O'Reilly Media, the year of the chatbot.[38] Microsoft executive Satya Nadella even claimed, in that same year, that bots were the new apps.[39] And in fact, chatbots have been quickly incorporated into the infrastructure of many chat applications, from mobile instant messengers to customer service windows on corporate websites. This begs the question: Where will the next Trojan horse appear?

It is time for policymakers and technologists to design products with the future in mind. In the short term, there are simple fixes to standing policy and media tools that should be implemented immediately. In early 2019, former FEC chair Ann Ravel, campaign finance researcher Hamsini Sridharan, and I authored an article that contained clear principles and achievable policies for countering digital deception. By "digital deception" we mean "the collection of opaque digital political advertising, malicious computational propaganda, and rampant disinformation spread by domestic and foreign actors that is destabilizing American democracy."[40]

The report details thirty-four unique policy suggestions, including the implementation of laws that have already been brought before the US Congress—like the Honest Ads Act, which would force social media companies to identify the prominence of political ads—but have stalled for one reason or another. Others are more novel and controversial, including amendments to Section 230 of the Communications Decency Act or, beyond politics, demonetizing the revenue streams of bad actors—the people who spread digital deception. We offer six principles that we believe will serve to protect democracy in the face of current and future malicious uses of digital technology: transparency, accountability, standards, coordination, adaptability, and inclusivity. These are the types of obvious legal remedies, and cogent values, that must guide both fixes to our current systems and preventative legislation covering future technology. These ideas are not intended to stifle innovation, but they do prioritize democratic freedoms over technological progress.

If technologists are worried that regulation will stifle innovation, then perhaps they are focused on the wrong thing. It is a hard fact that Google, Facebook, Microsoft, Amazon, and Apple have become monopolies in the search, social media, software, cloud storage, and hardware spaces. In my experience, turning the conversation to antitrust issues is the most effective way to get tech company employees to leave the room during a conference or closed-door meeting, but these conversations need to be had. We are talking about not only some of the most profitable companies on earth but companies that control a massive proportion of the world's data and news. The three statutes that promote fair competition in the United States are all over one hundred years old: the Sherman Antitrust Act of 1890, the Clayton Antitrust Act of 1914, and the Federal Trade Commission Act of 1914. It is time to revisit these statutes and begin applying them to new media companies.

Many have suggested that banning social media bots would deal with the problem of amplified propaganda online. In 2018, Senator Dianne Feinstein proposed the Bot Disclosure and Accountability Act of 2018, a bill to

> *protect the right of the American public under the First Amendment to the Constitution of the United States to receive news and information from disparate sources by regulating the use of automated software programs intended to impersonate or replicate human activity on social media.*

At the time of this writing, the bill was stalled in the Senate Commerce Committee.

Feinstein's bill is a step in the right direction, but it's overly broad. It may be realistic to ban social media bots on platforms yet to be created, but it's nearly impossible on a site like Twitter, where bots are infrastructural. To really get at the problem of amplification, lawmakers and technology firms should look to the problem of inorganic information flows. Using a variety of metrics—including network-, time-, and content-based markers—it is possible to track whether or not a political campaign is grassroots (from the people) or astroturf (from illegitimate actors). Coordinated armies of human accounts on social media can often be as successful as bots, and perhaps more successful, in seeding disinformation for political purposes. What if we limit the number of posts a person can make in a day? What about limits on posts about specific topics?

These last two proposals are more rhetorical than anything else, because they threaten free speech on the platforms. Such mechanisms for curbing propaganda might succeed in a country like Germany, which has had a history of prohibiting "harmful" speech since World War II, but they would have trouble achieving support in the United States. Both regulators and media companies, however, must find ways to curb hate speech. At present, efforts to enforce hate speech laws in online cases face serious obstacles. On many social media platforms—Twitter and Reddit, for instance—hate flows are anonymous. Moreover, many pundits and many regular people argue that even these laws—designed to protect minority groups—prohibit free speech because "hate speech" has no universal definition. In 2018, former ACLU president Nadine Strossen even released a book entitled *HATE: Why We Should Resist It with Free Speech, Not Censorship.*[41]

It is easier, though, to prohibit mediated content that either promotes violence or suppresses votes. In my research on computational propaganda, I have come across both types of content. I've cataloged posts containing both on Facebook, Twitter, YouTube, Reddit, and Instagram and even in Amazon reviews. Facebook worked to curb voter suppression efforts ahead of the 2018 US midterms, but such fragmented responses from social media firms—even the massive ones—do not effectively address the problem on their own.[42] Voter disenfranchisement campaigns stretch across multiple social media platforms and other websites, so any corporate attempts to self-regulate must be coordinated. Moreover, self-regulation by companies is not enough and will not lead to lasting change. Governments have to step up. Law enforcement agencies in the United States and elsewhere may have the means to put a stop to some types of online voter suppression, especially since many of these campaigns, experts suggest, are homegrown.[43] Violent content is marginally easier to prevent, due to clearer illegality or in the interest of facilitating parental protections, but even for government authorities, it is difficult to track on anonymous or encrypted websites and applications. The comments made on Gab by the Pittsburgh synagogue shooter just before he committed his heinous crime are one example among many.

Laws such as Marco Civil and GDPR attempt to curb the collection of user data, but we still have a long, long way to go in this regard. Social media companies and myriad other tech firms are actually data companies, after all. Most of the largest social networking platforms and search engines make money by selling access to users and their data. For example, in late 2018 the *New York Times* reported that for years Facebook had been giving more than 150 corporate "partners" special access to private user data—including users' private messages, without users' clear consent.[44] Companies claim, on the one hand, that they will not or cannot stop disinformation because they want to protect free speech, and on the other hand, that they are making moves to protect user privacy. At the same time, these firms are selling user data for profit. And yet some still argue, from *Forbes* to the budding conservative pundit Amelia Irvine, that the best path forward is either self-regulation or no regulation at all.[45]

Multibillion-dollar corporations, feckless governments, special interest groups, and technology investors are primarily to blame for the rise of computational propaganda. Google and Facebook built products without brakes and, at the risk of overextending the metaphor, without mirrors. Policymakers around the globe ignored the rise of digital deception, and many even profited from it. Faceless organizations built and launched online disinformation campaigns for profit, and others funded the construction of such efforts. Investors gave money to young entrepreneurs without considering what these start-ups were trying to build or whether it could be used to break the truth.

The global dilemma posed by computational propaganda is absolutely not a "technology users should pull themselves up by their bootstraps" situation. Regular people who go online or use social media certainly have a role to play in addressing the associated problems, but they alone did not create them.

Chapter Four
Artificial Intelligence: Rescue or Ruin?

Zuckerberg's MacGuffin

In April 2018, Mark Zuckerberg, founder and CEO of Facebook, appeared before congressional lawmakers in Washington, DC, to explain Facebook's data policies. The explicit purpose of requesting his testimony was to ask him to explain how his company made use of users' personal information: Did they protect it? Did they use it internally? Did they sell it? Did they sell it during elections? Did they sell it to foreign entities? But implicitly, and not so subtly, Zuckerberg was under the political microscope for Facebook's mishandling of user information during the 2016 US election.

He was also asked to account for a variety of other missteps associated with the online spread of "fake" news before, during, and after that particular contest. How did Cambridge Analytica get a hold of Facebook data on 50 million American users? How did the Russian government game Facebook to spread political conspiracy theories and polarize the voting public?

By the time Zuckerberg appeared in Washington, he had issued a very public mea culpa about Facebook's initially dismissive reaction to the charge of having spread disinformation and mishandled user data during the election. Directly after Trump was elected, Zuckerberg had cast shade on the allegations of political problems created by Facebook. "Personally

I think the idea that fake news on Facebook...influenced the election in any way is a pretty crazy idea," he said.[1] Days later he doubled down, saying that it was "extremely unlikely" that junk news stories had had any impact on voters or public opinion.[2] But between 2017 and the hearings in 2018, Zuckerberg engaged in what the *New York Times* called a "public apology tour."[3] He eventually apologized for dismissing the impact of fake news as "crazy", saying that he regretted his words.[4] It was as if he was going through the stages of grief—denial, anger, bargaining, and so on—in his response to allegations of Facebook's blunders. His appearance before Congress was his chance to set the record straight.

The news coverage and social media conversations after the event focused on Congress's lack of comprehension about basic functions of social media and the internet. Many of the questions the committee members posed to the Facebook CEO were as funny as they were frighteningly ignorant. Senator Lindsey Graham (Republican, South Carolina) asked Zuckerberg, "Is Twitter the same as what you do?" Senator Brian Schatz (Democrat, Hawaii) asked, "If I'm emailing in WhatsApp...does that inform advertisers?"[5] In an especially Magoo moment, seemingly not having been briefed by staffers on the topic of the meeting whatsoever, Republican senator Orin Hatch of Utah asked, "So, how do you sustain a business model in which users don't pay for your service?" "We sell ads," responded Zuckerberg, fighting to hide a smile. After all, it was the topic of ads and their relationship to user targeting that was the primary focus of the whole affair.

While I found some of Congress's questions notable for their obliviousness, it was something else that caught my ear. Zuckerberg mentioned artificial intelligence (AI) technology more than thirty times during his two-part testimony.[6] More than this, he suggested that AI was going to be the solution to the problem of digital disinformation, by providing computational programs that would combat the sheer volume of computational propaganda. He predicted that in the next decade AI would be the savior for the massive problems of scale Facebook and other companies face when dealing with the global spread of junk content and political manipulation online, problems that tech companies like Facebook created. He said, "Over the long term, building AI tools is going to be the scalable way to identify and root out most of this harmful content."[7] But Zuckerberg could provide only scant details on how this AI magic is going to happen. One expert, a Cornell Tech law professor, seized on this and told the *Washington Post*:

> *"AI is Zuckerberg's MacGuffin," said James Grimmelmann, using the film term for a mostly insignificant plot device that comes out of nowhere to move the story along. "It won't solve Facebook's problems, but it will solve Zuckerberg's: getting someone else to take responsibility."*[8]

AI tools will certainly play a role in fighting the flow of false information online. But Facebook's current AI tools cannot effectively manage false information on its platform alone.

Simple Bots to Smart Machines

Someday soon you log onto your favorite social media platform and notice that you've got a lot of new followers, more than you've ever had in a single day. Looking through the profiles of these people, you see that some are clearly in your professional field and some look like regular people. You wonder whether an idea you shared online the day before was particularly compelling. Or maybe these new followers are trying to build their professional network or trying to find a new job? With their normal profile pictures and past posts that seem very normal, there is nothing about these followers that is immediately suspicious.

In the coming days and weeks, your new followers begin liking things that you put on the platform. They share stories you post, they comment on things you say, and they even chat back and forth with you, mostly about work. You get to know these people and even begin to feel like they are your online friends. You talk shop with them, sharing thoughts about your career and the place where you work. In your effort to build social and career-oriented capital, you unwittingly share information that could be viewed as proprietary and that could be tied back to your company because of your visible employment there.

The problem is that these people are not your friends. They aren't even people. They are bots that were built using artificial intelligence software. They are purposefully constructed to gather information on people who work at a particular company. They learn all they can about you and other people like you, chatting with you and storing all the conversational details in a massive database. Eventually, those who run the accounts—people based at a firm specializing in corporate social listening—begin reporting what the AI bot accounts have learned. They share the things you have said with higher-ups at your own company. Maybe you get fired for breaking your nondisclosure agreement. You've been tricked, but your bosses are quick to tell you that what they did wasn't illegal. They were monitoring your communications on your company computer—which, incidentally, you weren't supposed to be using for social media purposes.

On the one hand, this hypothetical situation sounds far-fetched. It seems very unlikely that you would be effectively duped over social media by a "smart" bot account. On the other hand, it is strangely familiar. Monitoring employees' behavior on their work computers has become fairly common practice among employers. Beyond this, we all know now that people the world over have been duped by basic—not even AI-powered—bots. People have shared bot-driven articles and even unknowingly had arguments with bots on Twitter and Reddit. At the same time, AI is progressing at a fast clip and becoming cheaper. AI bots have been around for a long time, but now they are beginning to become available for broader public use. Politicians and people working to manipulate public opinion have taken notice. In the coming years, without action from governments and social media firms, AI bots will be used to target voters based on a panoply of information ranging from their employment history to where they live to what religion they practice.

This chapter is about AI and its role in the spread—and conversely, the containment—of computational propaganda. It's also about political bots that use AI to varying degrees. Most importantly, though, this chapter is about how smart technology systems and humans are already interacting and how they might interact in years to come. How is AI used to spread propaganda? How could it be used to fight it? How is it being used for other manipulative and disinformative purposes? What are technology companies and policymakers doing to address these uses? What aren't they doing? That being said, the Facebooks and Googles of the world could be doing all that they can to limit misuses of their technology, but it would still not be enough.

Let me be clear: these powerful companies are very far from actually doing all that they could, technologically or otherwise. At the end of the day, technology firms are still focused on two things before all others: profit and growth. There is nothing inherently wrong with the desire to make money or to scale a company, but there is a problem when the bottom line comes before our collective freedom.

At the same time firms are trying to curb the impact of disinformation they are also building and deploying new tools that threaten to empower it. For instance, SalesForce and Microsoft have both built software products that summarize news articles using AI.[9] Google has built a machine learning bot that can write Wikipedia articles.[10] The potential for misuses of these products is massive, and the repercussions of getting these kinds of information wrong could be deadly. So much for tech firms not being "arbiters of the truth." Not only do these new technologies help people get around the paywalls of newspapers trying to stay afloat, but they also expose the continued likelihood of tech firms prioritizing, mistranslating, or misinterpreting current events. These new products provide, as my keen-minded editor put it, "just one more layer of simulacrum in the game."

This may sound bleak, but technological fixes on their own will not cure the complex ills that we face as a global public. We need to address and mend divisions in society, acknowledging that most of day-to-day life still happens offline. We need to invest in social programs that empower those who have been most disadvantaged in systems of power: ethnic and religious minority groups, people of color, women, and people with disabilities. Regardless of your political beliefs, these groups are absolutely integral to the functioning of democracy. We need to leverage the power of society, of difference and equity, to make changes to technological and social systems that take advantage of those who lack a voice in mainstream politics and culture.

The uses of the technologies that are rocking society in the United States are even more damaging in the developing world and for those who are most subject to underlying social and political constraints—minority and disadvantaged groups. In fact, it is these constraints that allow for certain manipulative uses of social media and emerging technology in the first place. Traditional redlining—the denial of services due to race or ethnicity—is still happening the world over, and now, with the help of big tech, it's supplemented by digital redlining: the filtering of certain content online by race or ethnicity. One result is that a small group of tech company executives are among the richest people in the world.

According to a 2016 *Business Insider* article, the thirteen richest people in tech hold as much as $450 billion in wealth.[11] According to the World Bank, that is over $100 billion more than the gross domestic product (GDP) of the banking megalopolis city-state of Singapore, which the *Economist* Intelligence Unit has ranked as the world's most expensive city for over five years in a row, from 2013 to 2018.[12] It is presumptuous to assume that this small group of billionaires will put democracy and the truth before profit and growth, or the needs of the less fortunate before the interests of the rich.

We need systems that act in favor of democracy and equity, regardless of the desire of a wealthy few.

Some might argue that societal fixes are not the responsibility of technology firms. I disagree. Corporate social responsibility (CSR), both inside and outside of the technology sector, has been gaining traction in the business community for years now. Our ideas about what CSR entails have been evolving. Scott Shackelford, an Indiana University professor of business law and ethics, makes a convincing argument that privacy protection, for instance, should be a part of CSR.[13] He writes that, "if Facebook declared its support for both privacy and security as inalienable human rights akin to internet access, that could help the company get started, before policymakers in the US and around the world step up to have their say."

Policymakers, tech companies, and society itself all have their share of responsibility in rebuilding democracy for the digital age. Bearing the greatest responsibility to fix what they have broken (or ignored until it was broken) are tech companies, which built the tools, and policymakers, who are in charge of regulating the media. It will be society that holds people accountable. Contacting elected leaders to put serious pressure on them to act in the best interest of society rather than in the interest of corporate lobbyists still matters a great deal. Although there is no single route to solving the problems posed by manipulative uses of technology, there is still power in organized society. We need a systematic way to respond to computational propaganda and disinformation, not a magic technological bullet. AI may be a groundbreaking and ever-changing technology, but it will not solve the problem alone, no matter what Zuckerberg argues.

A User Issue?

Consider the following situation. Individuals use social media every day, sometimes for many hours. They log onto Facebook to find out what their family and friends are up to or to engage in discussions about their favorite hobbies or political issues on a group page. They post photos of their lives on Instagram and regularly scroll through images related to style, news, travel, or fitness using the platform's search and recommendation mechanisms. They use Twitter to send messages about current events and to follow journalists who report on global affairs. Like their peers, these users spend more time on social media than they do tuning in to many types of traditional media like TV or radio. These platforms are the places they visit to connect with society, with public life.

In each of these experiences, users are shown advertisements featuring, for instance, the latest toys, furniture, and entertainment technology. Many of these ads seem to be correlated with recent searches they did for similar products, using other websites. Users are also shown political advertisements. Some are smear ads, targeting particular politicians, while others are endorsements of certain causes or attacks on other causes. Some of the ads link to news articles that purport to offer facts as to why a certain candidate or cause is good or bad. Closer inspection of the articles, however, reveals that they are not only subjective but totally nonfactual. Full of rumors and opinion, these pieces are much more propaganda than anything akin to news. And strangely, the ads that lead to these articles do not say who paid for them. Users can't tell who wrote the fake articles, or why. They cannot see who is behind the attacks—no name of a super PAC or proxy organization, however obscure, is provided. Not only do users not have all the information about these ads, but they probably don't even know it. Meanwhile, users see these ads nearly every day, across multiple platforms.

This very scenario has been experienced by millions, perhaps billions, of people who used social media in the years preceding our current moment. They were scammed by political groups hoping to get them to donate money, to vote a certain way, or to believe a particular rumor. They were targeted by foreign governments that never had to register or provide details to the social media firms that allowed them to buy ads and start phony political group pages. Users had little to no information on who was targeting them with particular types of information, and no insight into how the algorithms underlying social media news feeds or trends were prioritizing certain kinds of information. They were provided with opaque (at best) terms of services detailing the sale of their intimate personal data to the very companies and groups working to manipulate them.

It has become popular in expert circles, especially when tech leaders and politicians are doing the talking, to say that online disinformation is "a user issue." When leaders from either realm say this, the buck is being passed: those most responsible for the problem are arguing that the solution is to make users—the regular people duped by the numerous unfair and illicit practices just detailed—fix the issue. They only have to give people a bit more information, they say—an AI-driven plug-in that detects bots, a machine learning fake news-o-meter, or a deep learning system for down-ranking content. They are arguing, in other words, that users need to self-correct. But regular people are not only beyond overloaded with information—they are victims of deceit. They are victims of the severe lack of foresight of social media firms, the dupes of digital advertisers, and prey for malicious political groups.

The "it's a user issue" argument is flawed and problematic for a variety of reasons. It puts the burden of responsibility on ordinary citizens who are being simultaneously taken advantage of by vast corporate entities, who make a living off their data and attention, and failed by Luddite politicians like Senators Graham and Hatch. Regular people's personal data is being bought and sold by social media firms to faceless political campaigns and predatory groups. They are then being targeted, using their own data, with online political ads often laden with false information. Wherever they go across the web, they can find little or no information on who is actually behind these ads: no buyer names are given and no responsibility is taken. Self-regulation is not enough. Facebook and Google cannot simply implement their own "honest ads" policies and then assert that they are solving the problem. They are not even coming close to doing their part in solving it.

In the end, very real decisions (or often a lack of decisions) about the design and management of social media and the broader internet have enabled the digital spread of junk news and disinformation. YouTube, Facebook, and Twitter curate and package information in very real ways. Since they decide what users see and when they see it, they are responsible for that information. At the same time, a lack of willingness, or ability, to govern the use of the internet during elections has allowed digital deception to flourish. It is past time for the Federal Election Commission and election oversight groups in other countries to sensibly regulate digital political communication.

Dumb Bots

Anytime you log onto Twitter and look at a popular post, you are likely to find bot accounts liking or commenting on it, despite changes Twitter has made to try to stop the broad use of social automation on the platform. Clicking through to these accounts, you can see that they have tweeted many times, often in a short time span. Sometimes the content of their posts is oriented toward spam—selling junk or spreading digital viruses. Other accounts, especially bot accounts that post garbled vitriol in response to particular news articles or political statements, are entirely political. Most such bots aren't AI-empowered automatons but just dumb bots, built to spread the same trolling content over and over again in response to other accounts using a particular hashtag or tweeting about a particular account.

It's easy to assume that both computational propaganda and the current wave of disinformation are driven by advanced computer science. I've talked to many regular people who believe that ML- or AI-driven algorithms allow tools like political bots to learn from their surroundings and interact with people in a sophisticated manner. After the 2016 election, pundits and journalists fueled this fire. My own research has actually been quoted in articles that suggest that our robot overlords are already here. There are a slew of such pieces.

In 2017 the news start-up *Scout* wrote an extremely provocative piece about the rise of a "weaponized AI propaganda machine."[14] A few months later the research-oriented reporting outlet *The Conversation* put out another such piece, later reprinted in the online UK newspaper the *Independent*, that claimed that "artificial intelligence conquered democracy."[15] Around the same time the author of this piece, Oxford PhD student Vyacheslav Polonski, wrote a similar post for the Council on Foreign Relations' *Net Politics* blog. This time he argued that "artificial intelligence has the power to destroy or save democracy."[16]

The fact is, though, that complex mechanisms like artificial intelligence have played little role in computational propaganda campaigns to date. The political bot campaigns that my collaborators and I have cataloged have been automated, but rarely intelligent. All the evidence I've seen on Cambridge Analytica suggests that the firm never launched the "psychographic" AI/ML marketing tools they claimed to possess during the 2016 US election.[17] At Oxford we looked into how and whether Twitter bots were used during the Brexit campaign.[18] We found that a lot of political bots were being used to spread messages about the Leave campaign. The vast majority of the automated accounts we found were very simple, made to retweet content or generate noise, not to be functionally conversational—they did not harness AI. In the Institute for the Future's Digital Intelligence Lab's examination of political bot usage during events in places like Turkey, Venezuela, and Ecuador, we saw bots programmed to target journalists and civil society leaders with repetitive harassment. The automated social media accounts were not artificially intelligent. They simply overwhelmed their victims with cascades of repetitive automated hate.

Nearly all of the computational propaganda campaigns I've studied over the years have been wielded quite bluntly. During events in which, researchers now believe, political bots and disinformation played a key role—the Brexit referendum, the Trump–Clinton contest in 2016, the Crimea crisis—smart AI tools have played little to no role in manipulating political conversation; not even marginally conversational AI tools have been employed. Online communication during these events was altered by rudimentary bots that had been built simply to boost likes and follows, to spread links, to game trends, or to troll opposition. It was gamed by small groups of human users who understood the magic of memes and virality, of seeding conspiracy ideas online and watching them grow. Conversations were blocked by basic bot-generated spam and noise, purposefully attached to particular hashtags in order to demobilize conversations. Links to news articles that showed a politician in a particular light were hyped by fake or proxy accounts made to post and repost the same junk over and over and over. Manipulative human groups sowed junk news in prominent Facebook groups and left it there to fester and spread.

This is not to suggest that the people using these simple bot armies or other inorganic information operation tactics were not successful in their efforts. In fact, these individuals and groups realized that they did not need sophisticated conversational AI to hack public opinion. AI software tools are not only expensive but, until recently, hard to come by. The propagandists in question could game the truth by using very rudimentary political bots to overwhelm the trending algorithms on social media sites. They figured out that social media sites, at least early on, had fairly basic number-driven metrics for determining popularity. Even people without many resources could make an article or a political meme look popular when it flat out was not. This said, it has historically been governments and other powerful political groups that have built the largest and most persistent political bot armies—sophisticated technology or not.

When moneyed digital propagandists have needed to send more complicated or nuanced messages online, they have paid people to do it. For instance, research from Harvard's Gary King and Stanford's Jennifer Pan suggests that the 50-Cent Army in China, known for spreading effusively positive information about the country's Communist Party, makes use of tens of thousands of people—not bots—to spread political content. In the United States, journalists and researchers have discovered that small groups of far-right activists have been successful in getting much larger groups of people on Twitter to spread disinformation.[19] In these cases, thousands of unsuspecting folks in, say, Ohio and Illinois are tricked into spreading disinformative memes generated by white nationalists or anti-vaxxers. These examples in both China and the United States can be thought of as similar to astroturf efforts—in China simply because the 50-Cent Army's efforts are driven by the government, and in the United States because these efforts by fringe political groups involve tricking politically moderate American citizens into spreading partisan disinformation.

What is especially worrying is that Twitter and Facebook have a very hard time detecting and deleting human-driven propaganda campaigns. Computational propaganda campaigns that use cyborg accounts, which run via a combination of human and bot labor, are on the rise. Networks of social media users that made use of both human labor and automated software existed when I first began studying these problems. Accounts that my team at Oxford deemed Russian cyborg accounts attacked journalists and readers in the comments section of *Guardian* articles about Malaysian Airlines Flight 17 and the Crimea crisis.[20]

In Brazil's 2014 general election, people who ran these hybrid propaganda accounts spoke to news outlets, including the BBC, about their work. These paid "activators" told the BBC's Brazil outlet that they "would praise whichever political candidates they were being paid to support, attack their opponents and sometimes join forces with other fake accounts to create trending topics."[21] In fact, said one of these activators, "either we would win [debates] through sheer volume, because we were posting so much more than the general public could counter-argue. Or we would manage to encourage real people—real activists to fight our fight for us."

The Era of AI Bots

Because better communication technology is becoming cheaper and more ubiquitous, the era of computational propaganda spread by just simple bots or coordinated groups of people is ending. We are moving toward a period of advanced digital propaganda. Cyborg accounts are being used to spread computational propaganda more than ever before. Increasingly smart technology, political bots, and other tools that do make use of advanced machine learning and deep learning programs are beginning to play a role in the spread of computational propaganda and disinformation. Regardless of the differences between its various parts, as AI technology becomes more available, automated conversational agents online will become more and more humanlike. There will be a fine line between an apparently human account online and an apparently botlike one.

AI and Its Parts

In the world of computing, AI refers to any computer program that makes "smart" decisions to achieve a particular goal based on the data in its environment. John McCarthy, an early computer and cognitive scientist who coined the term "artificial intelligence" and pioneered early research on the topic, called AI "the science and engineering of making intelligent machines." Machine learning (ML) is a subset of AI in which a program is built to alter itself, without explicit human intervention or further human coding, when it encounters more data. These algorithms are focused on "optimization"—learning from information sets in order to maximize task efficiency and minimize error. Deep learning (DL), in turn, is a category of machine learning that tends to encapsulate the use of artificial neural networks, which are modeled on the human brain, and deep reinforcement learning. One way of explaining deep learning is to say that it is a more accurate and complex version of machine learning that is especially useful in parsing less structured data.

Now imagine a political bot that makes use of something like machine learning. Such a program would be able to learn from conversations it has with people over social media and would constantly get better at having them. If, for instance, the bot was built to convince people to share an article arguing against global warming, it would use data from its interactions to figure out what tactics work and which do not. It would then amend its communication accordingly, until it achieved optimum success in getting others to share the article.

Conversational machine learning bots have existed, relatively speaking, for a long time. But until recently they required too many resources—not just money but also serious time and intellectual labor—to deploy in numbers. This is often still the case, but not for long. New chatbot systems can learn from data, identify patterns, and make decisions with minimal human intervention. Moreover, they are becoming cheaper. How long will it be before political bots are actually the "intelligent" actors that some thought swayed the 2016 US election rather than the blunt instruments of control that were actually used?

At present, most people are not thinking about the influence of smart bots in politics. We are still reeling from issues relating to Russia's online machinations during the 2016 US election, the death of tens of thousands after military-led Facebook propaganda campaigns in Myanmar, and the global rise of "fake news." These are sociopolitical uses of social media sites using tools that networks like Facebook and Twitter designed and offer for public use and for sale. Using advertising on these sites to spread propaganda is not hacking—it is business as usual. Manipulating social or interest-driven groups to polarize and enrage them, using Facebook's group page function, doesn't require AI.

Relatively few world leaders are paying attention to the current profusion of signals that foreshadow how technologically enhanced propaganda will progress. The 2018 midterm election in the United States showed us that already marginalized social groups, which are essential to the functioning of a true democracy, are now primary targets of online disinformation and political harassment.[22] Recent contests in Egypt, Hungary, India, Iraq, Mexico, Russia, Sweden, and Turkey have revealed that digital disinformation, online conspiracy, and automated amplification continue to play significant roles in political communication. These problems show little sign of abating, and neither legislation nor self-regulatory moves from social media firms have effectively or comprehensively addressed the issue.

Of course, computational propaganda and digital disinformation are complex problems. Like the web, the tools and tactics that groups use to spread truth-busting content are constantly changing and progressing with technological innovation. Moreover, both new and old technology companies, including social media platforms and chat applications, continue to grow at a fast clip. Many of them are so focused on innovation, scale, and novelty that they continue to fail to heed the warning signs. Today these firms and other groups working to combat computational propaganda address information operations and the problems associated with them in a post-hoc manner, well after the manipulation has occurred and the tactics of propagandists have been cemented into a more robust strategy. It is time for them to put a serious amount of their innovative energy toward dealing with disinformation campaigns in real time and, even more importantly, toward planning for future misuses of the technology they build.

Forgetting Ethical Design

A group of software engineers who work for an augmented reality (AR) social media company called YourFace are gathered in a boardroom. On the AR YourFace platform, users wearing a cheap set of digital-connected glasses can see all sorts of virtual content mapped onto their normal everyday space. It's like Pokémon GO, but bigger, more social, and more about sharing interesting factoids about real-world locales—an AR

Wikipedia. The public is in love with the new platform, which has gained a billion users in its first year.

However, the engineers concede, there is a small problem—certainly not their fault, but a problem regardless. The open API on YourFace, which allows anyone to launch new virtual content onto the world viewed through the glasses, has facilitated the spread of hate speech and harassment on the platform. Information-scapes that were once mostly objective now feature racist, sexist, and homophobic images and content.

The engineers decide that they have found the perfect solution. They will build an algorithm that detects the hate speech and removes it. They can keep the open API while still protecting vulnerable groups. They train a machine learning algorithm and deploy it two months later. It is a phenomenal failure. Not only does it not catch much, if any, hate speech, but it seems to be totally out of touch with the actual experiences of anyone who might experience this type of harassment. It turns out that the group of engineers, all white and all male, were the ones who trained the AI algorithm. Unsurprisingly, they built a piece of software that was completely flawed. They had technical know-how, but they had little social or ethical awareness.

This scenario, while fictitious, is not far off the mark as a description of what has happened behind closed doors at social media firms in recent years. Much of the propaganda, vitriol, and trolling that have spread in the last decade have been enabled by the failure of technology firms to design their tools with either human rights or democracy in mind. Facebook and YouTube did not set out to build ethical, equitable technology that would consider the problems that diverse groups of users might experience in years to come. And they still haven't truly figured out how to respond to the problem of computational propaganda. They simply did not plan for the political co-optation of their platforms because they were captivated by the constant rhetoric from top tech executives about their platforms saving democracy simply by connecting the world.

A former Facebook employee with close knowledge of that company's efforts to curb what it calls "information operations" told me that Facebook continues to scramble to respond to these failures and oversights. She painted a bleak view of a disorganized and disenchanted company that hires security contractors and disinformation experts left and right to see if maybe those on the outside have any of the answers. Her implication, simply put, was that the social network's response has been anything but systematic. She, like Zuckerberg, mentioned AI tools as the only real way to address the ever-growing problem of political manipulation on the platform—but she also said that it was clear that the company needed to focus more on the social aspect of the problem.

One concerning and confusing thing about Mark Zuckerberg and Facebook's AI defense is the fact that elements of AI—and of ML—seem poised to play a significant role in future efforts to spread false information online during elections and crises the world over. Essentially, those offering AI as a solution are suggesting that we fight fire with fire. Will this result in a technological arms race? And if it does, where will it stop? Most of the political bot makers I've spoken to over the last five years have mentioned machine learning in some way, shape, or form. Again, they usually talk about it in hypothetical terms, explaining that most of the ways one can use social media bots to manipulate public opinion are somewhat technologically unsophisticated—that is, they do not require AI or ML to achieve their ends. Other bot makers, though, have already built and deployed small groups of bots encoded with various ML capacities.

For instance, a software engineer named Mercer told me about a bot he built that attempted to have conversations with politicians and pundits. The end goal was for the bot to get any form of interaction with powerful political figures. Having these political figures retweet the bot's content—thus validating that content—was the gold standard, because they were undoubtedly spreading the automated content to many more people. Mercer's bot was successful at engaging with all sorts of big names. It was designed to learn from its conversations, from both failures and successes. It tested out various conversational tactics—asking a question in a particular way, bringing up a recent event, praising the person in question—in attempts to get attention and spark conversation. It also scraped basic information on the people targeted for communication by its maker and used details it thought might entice them into conversation. What better way to get someone to talk to you or retweet you than to talk to them about what they like or to stroke their ego?

If you examined this particular bot account closely, or if you engaged in conversation with it for more than a couple of lines, you would definitely figure out that something was odd. There was some degree of an "uncanny valley" effect—the account seemed sort of human, but strange little things were off.[23] The account spelled words wrong, messaged random content, and sometimes shared messages in other languages. These mistakes might be chalked up to coding errors by the builder, but my own experience interacting with and studying these sorts of accounts suggests that it is not uncommon for present-day ML social media bots to have a similar tell. Moreover, ML bots like Mercer's do not yet—and may never truly—have the ability to parse qualities like humor, sadness, or sarcasm the way a human can. The field of sentiment analysis is changing this by working to systematically categorize these types of subjective information, but this is a quantitative attempt to capture and explain something that is thoroughly qualitative—in other words, something very human. It is a challenging task, to say the least.

Although we cannot count on AI and its various subfields alone to save us from the problems generated by particular human uses of technology, Zuckerberg is at least partially correct in that these tools do have a part to play. The most effective response to the current techno-facilitated twisting of reality will combine the social and the technical, and it will encompass short-term, medium-term, and long-term solutions. This response will include not only new tools but also novel forms of media literacy.

We can use AI—and tools like VR, AR, video, and voice emulation systems—to bolster efforts to build cognitive immunity around the globe, but we cannot rely on these devices alone. Even the computational tools we build with the intention of aiding democracy should be closely scrutinized for potential ethical problems before they are launched. Diverse groups must be represented in efforts to vet socially focused AI tools before they are launched into use by the broader public. Kevin Kelly, founding executive editor of *Wired* magazine, has suggested that technology generates as many new problems as new solutions. Most of the problems we will have in the future, he argues, are going to be caused by the technologies that look like solutions to us today.[24]

The Beginnings of AI Propaganda

There are signals that AI-enabled computational propaganda and disinformation are beginning to be used, as well as evidence of such content from the bot makers I've talked to in my research. Since Twitter launched in the mid-2000s, people have been testing and refining conversational AI bots. But the use of these bots in politics has been limited until recently. Oxford researcher Lisa-Maria Neudert notes in an article for the *MIT Technology Review* that the use of smarter chatbots over social media during political events is close at hand.[25] "Rather than broadcasting propaganda to everyone, these bots will direct their activity at influential people or political dissidents," she writes. "They'll attack individuals with scripted hate speech, overwhelm them with spam, or get their accounts shut down by reporting their content as abusive."

Hackers and other groups have already begun testing the effectiveness of "weaponized" AI bots over social media. According to a 2017 piece from *Gizmodo*:

> *Last year, two data scientists from security firm ZeroFOX conducted an experiment to see who was better at getting Twitter users to click*

on malicious links, humans or an artificial intelligence. The researchers taught an AI to study the behavior of social network users, and then design and implement its own phishing bait. In tests, the artificial hacker was substantially better than its human competitors, composing and distributing more phishing tweets than humans, and with a substantially better conversion rate.[26]

The bots were more effective in this experiment—and will be more effective in future applications—because they can operate at scale. This capacity, combined with the ability of some ML-trained bots to personalize the information they direct at victims, makes for a very potent tool for manipulating public opinion on a granular level. And while the *Gizmodo* example was a single experiment, it is likely that AI will soon be used more broadly for nefarious ends—if it's not happening already under the radar. The article points to a poll held by the cybersecurity company Cylance during the 2017 Black Hat USA hacking conference: when asked whether they thought AI would be weaponized in the next year, 62 percent of attendees said yes.

Manipulative and problematic AI content is not limited to being spread by ML-enabled political bots. Nor are problematic uses or designs of technology being generated only by social media firms. Researchers have pointed out that ML can be tainted by poison attacks—malicious actors influencing "training data" in order to change the results of a given algorithm—before the machine is even made public. Training data is the material given by a programmer or engineer to an intelligent machine to help it become conversant in a given discipline.[27] This type of tainted code could then be used for all manner of purposes, well beyond chatbots. Facial recognition software, often trained using ML, can be built with seriously racist biases.

Internet entrepreneur Kalev Leetaru argues that although we have not yet seen large AI-powered manipulative botnets, they are on their way.[28]

He also suggests that the first AI bot-driven attacks may not be socially facing: "Perhaps the first incarnation of autonomous offensive cyber-warfare will come in the form of intelligent DDoS attacks." In a DDoS, or a distributed denial of service attack, multiple automated (bot) systems flood and consequently shut down targeted web servers. Such offensives have been used with great effect by governments around the world to shut off the sites of their adversaries. But Leetaru takes this idea further:

> *Now, imagine for a moment that you handed that botnet over to the control of a deep learning system and gave that AI algorithm complete control over every knob and dial of that botnet. You also give it live feeds of global internet status information from major cybersecurity and monitoring vendors around the world so it can observe second-by-second how the victim and the rest of the internet at large is responding to the attack. Perhaps this all comes after you've had the algorithm spend several weeks monitoring the target in exquisite detail to understand the totality and nuance of its traffic patterns and behaviors and burrow its way through its outer layers of defenses.*

The implication of this scenario, he argues, is that we could end up with something close to Skynet—an unthinkably powerful cyber-weapon that can seize control away from human intervention and defy its attempts to respond to a breach or attack. And while it is conjecture, there is some weight behind the idea that DL or ML could be used to enhance socially oriented bots—and perhaps already are.

Microsoft's own failed AI bot experiment, called "Tay," showed that smart bots can be built to communicate with people on social media. It also revealed that these programs can be easily manipulated by other users for problematic and hurtful ends. Tay began repeating racist and anti-Semitic content shortly after it was launched on the platform. The automated profile, meant to talk and act like a teenage girl, was quickly gamed by online trolls who figured out that it had a "repeat-after-me" function. Microsoft took the bot down, quietly relaunched it days later, and then quickly took it down again after it continued to spew racist garbage.

The issue with using AI-driven bots for computational propaganda campaigns is that they are expensive and difficult to create, and thus hard to operate at scale. While an army of bots to like Twitter posts is relatively easy to build and launch, an army of AI bots that can learn from and talk with people is not. Simple bots can be used to manipulate public opinion outside of politics. Recent evidence suggests that Russian government actors have used social bots that spread anti-vaccine information, disinformative content about genetically modified organisms (GMOs), and junk science purporting to debunk global warming.[29] Governments, including Saudi Arabia and Russia, have even taken over the real accounts of real people and used them to spread both automated and human-derived propaganda. According to blogger Marc Owen Jones, who studies technology and the Middle East, the Saudis hacked and took over the verified Twitter accounts of a deceased American meteorologist and several others in order to spread pro-regime propaganda.[30]

Fighting Fire with Fire

So how can AI or automated bot technology be leveraged to fight the manipulation of public opinion online rather than further it? Can we fight AI with AI? It is certainly true that AI and ML can be useful in helping to detect computational propaganda. The Observatory on Social Media (OSoMe) at Indiana University has built public tools that harness ML to detect bots. Its website describes the functions of one of the observatory's best-known ML ensembles:

> *Botometer is a machine learning algorithm trained to classify a [Twitter] account as bot or human based on tens of thousands of labeled examples. When you check an account, your browser fetches*

> *its public profile and hundreds of its public tweets and mentions using the Twitter API. This data is passed to the Botometer API, which extracts about 1,200 features to characterize the account's profile, friends, social network structure, temporal activity patterns, language, and sentiment. Finally, the features are used by various machine learning models to compute the bot scores.*[31]

This particular tool shows promise in scaling political bot detection. Other AI and ML tools will undoubtedly be similarly helpful in detecting false news and information on social media. Some already exist, and some companies, including Facebook and Twitter, are using them. According to a 2018 blog post on fighting false news, Facebook product manager Tessa Lyons said that "machine learning helps us identify duplicates of debunked stories. For example, a fact-checker in France debunked the claim that you can save a person having a stroke by using a needle to prick their finger and draw blood. This allowed us to identify over 20 domains and over 1,400 links spreading that same claim."[32] In such cases, social media firms can harness ML to pick up, and even verify, fact-checks from around the globe and use these evidence-driven corrections to flag bogus content.

There is a big debate in the academic community, however, as to whether passively identifying potentially false information for social media users is actually effective. This argument is compelling whether or not AI or ML is used to detect junk content. Some researchers suggest in fact that fact-checking efforts on- and offline do not work very effectively in their current form.[33] In early 2019, Snopes, which had partnered with Facebook in such corrective efforts, broke off the relationship. In a candid interview with *Poynter*, Vinny Green, Snopes's vice president of operations, said, "It doesn't seem like we're striving to make third-party fact checking more practical for publishers—it seems like we're striving to make it easier for Facebook."[34] Green's assessment and comments by other third-party groups that have worked with the social media firm highlight the fact that organizations like Facebook continue to rely on small, usually nonprofit organizations to vet content. Potentially false articles or videos are often passed to these groups with no background information on how or why they were flagged by Facebook.

Such efforts aren't geared toward helping news organizations vet the heaps of content or leads they receive each day to help underresourced reporters do better work. Rather, they aid a multibillion-dollar company in keeping its own house clean in a post-hoc fashion. Foisting the work upon organizations like Snopes and news organizations like the AP helps Facebook avoid notice that it is in fact very much in the media business. It is time for Facebook to take responsibility internally for fact-checking, rather than passing off the task of verifying or debunking news reports to other groups. Facebook and other social media companies must also stop relying on fact-checks after the fact—that is, after a false article has gone viral. These companies need to generate some kind of early warning system for computational propaganda.

There are areas where Facebook is progressing in this regard. In an interview with *BuzzFeed*, Facebook's Lyons reported that "copycat hoaxes have been an increasing trend in 2017 and also into 2018"—copycat hoaxes being the use by various groups of similar strategies and similar content to spread rumors and false information. According to Lyons, "using machine learning we're able to identify and demote duplicates of articles that were rated false by fact checkers." Again, though, once the machine-learning algorithm detects a problem, the burden of moderation efforts is shifted to external fact-checkers who already face a barrage of junk and vitriol every day and also are battling tight budgets—their organizations being strapped for cash in large part owing to social media's effect on the news market.

There is one key question when it comes to exploring how social media firms like Facebook might best work to leverage AI technology to combat troublesome automation, disinformation, and political harassment: Can companies effectively combine this advanced technology with the efforts of their own human laborers to catch and debunk false information across myriad cultural contexts before it goes viral? Facebook, Google, and other companies like them employ people to find and take down content that contains violence or information from terrorist groups. They are much less zealous, however, in their efforts to get rid of disinformation. The plethora of different contexts in which false information flows online—everywhere from an election in India to a major sporting event in South Africa—makes it tricky for AI to operate on its own, absent human knowledge. But in the coming months and years it will take hordes of people across the world to effectively vet the massive amounts of content in the countless circumstances that will arise.

Another consideration is the serious psychological effect on the many contractors employed by tech firms to moderate content when they are dealing with the horrendous content that makes it onto these platforms.[35] In 2019 *The Verge* conducted a series of interviews with contract moderators employed by the company Cognizant to vet content on Facebook.[36] According to the article, "secrecy…insulates Cognizant and Facebook from criticism about their working conditions…They are pressured not to discuss the emotional toll that their job takes on them, even with loved ones, leading to increased feelings of isolation and anxiety." The article goes on to say that "people develop severe anxiety while still in training, and continue to struggle with trauma symptoms long after they leave," and that "the counseling that Cognizant offers them ends the moment they quit—or are simply let go." These contractors do not receive the comprehensive compensation packages paid to full-time Facebook employees. They seem to be hidden away by design.

There simply is no easy fix to the problem of getting rid of computational propaganda on social media. It is the companies' responsibility, though, to find a path forward. To date, investigative reports like the piece from *The Verge* have revealed that companies like Facebook have been far more focused on optics—on PR—than on regulating the flow of computational propaganda or, say, graphic content. According to that piece, the company has allegedly spent more time celebrating its efforts to get rid of particular pieces of vitriol or violence than on systematically overhauling its moderation process. *Salon*, in a response to Facebook's declaration that it would shift to become a "privacy focused" company, asks: "Is this Zuckerberg's solemn swear, or a mere public relations move to rebrand a company not exactly known for being trustworthy on privacy?"[37]

Beyond Fact-Checking

It will be some combination of human labor and AI that eventually succeeds in combating computational propaganda, but how this will happen is simply not clear. As reporter James Vincent puts it, "Ultimately, it's schemes like these—which involve human fact-checkers—that are most effective in identifying fake news. But AI can still provide a useful backup."[38] Other companies and organizations make use of similarly hybrid, or cyborg, models. In a 2018 article, the Institute of Electrical and Electronics Engineers (IEEE), a major professional association in the technology and computer science space, echoed the idea that hybrid models are the way forward: "Facebook, Google, and smaller tech companies are now using machine learning to flag misinformation—but automated systems aren't reliable enough on their own."[39] The article points to Full Fact, a UK-based fact-checking nonprofit, as an example of one entity that has found success in using an AI–human model.

But AI-enhanced fact-checking is only one route forward. Machine learning and deep learning, in concert with human workers, can be leveraged in several other ways to combat computational propaganda, disinformation, and political harassment. Jigsaw, the Google-based technology incubator (a term for entities that specialize in the early stage development of digital tools) where I served a one-year term as a research fellow, designed and built an AI-based tool called Perspective to combat online trolling and hate speech. This tool—which I myself didn't work on—presents another unique, though controversial, path forward in combating online vitriol. The technology is an API that allows developers to use the incubator's anti-troll toolkit to automatically detect toxic language. It is controversial because it not only runs the risk of getting rid of false positives—posts that do not actually contain trolling or abuse—but also moderates speech. According to a *Wired* article, the tool was trained using machine learning, but it is important to note that any such tool is also trained using human input—as with the fictitious YourFace algorithm dreamt up at the beginning of this chapter. As we know all too well, humans have their own slew of biases. So could a tool built to detect racist or hateful language actually fail to do so because of poor or flawed training?

In 2016 Facebook launched Deeptext, an AI tool similar to Google's Perspective. According to a 2017 article on the algorithmic arbiter, it helped the company delete over 60,000 hateful posts a week.[40] Company representatives admitted, however, that the tool still relied on a large pool of human moderators to get rid of harmful content. Twitter, meanwhile, finally made moves at the end of 2017 to work more carefully to ban similarly threatening or violent posts on the site.[41] But while the company has started curbing this problematic material—and is also deleting swarms of political bot accounts—its responses amount to "guidelines" and internal policies and do not include clear indications of how it is functionally detecting and deleting accounts. And in spite of all this, my research collaborators and I continue to find manipulative botnets on Twitter nearly every month.

AI tools built to monitor hate speech and propaganda sometimes miss the mark. In mid-2018 Facebook's detection tools managed to label a portion of the US Declaration of Independence as hate speech; the company apologized and reinstated the deleted post a day later. The *Liberty Country Vindicator*, a local newspaper based in the small town of Liberty, Texas, had posted excerpts from the Declaration on its Facebook page. Facebook promptly responded, singling out the portion of the venerated document that contains the phrase "merciless indian savages." Although the algorithm correctly caught the problematic and hateful phrase, it was not able to contextualize it historically within the larger body of the Declaration, and critics argued that its censorship was a potential violation of free speech.[42] This was not the first time that using AI led to the banning or deletion of historical or artistic content. In another case, an Instagram account featuring LGBTQ history was deleted for quoting a well-known poem from Zoe Leonard that contained the line "I want a dyke for president."[43]

These cases underscore the fact that AI, while useful as scaffolding for human efforts to find and moderate content, can get it wrong. . .a lot. Not only can such technology misidentify literary content, but "smart" algorithmically driven tools can be encoded with racism, sexism, and homophobia. For instance, US representative Elijah Cummings, chairman of the investigation-oriented House Committee on Oversight and Reform, had noted that AI facial recognition software can contain serious bias. In response to the use by US law enforcement agencies of these "biometric" tools to make arrests, the representative said, "If you're black, you're more likely to be subjected to this technology and the technology is more likely to be wrong."[44] Politician Alexandria Ocasio-Cortez, adored by many on the left and reviled by many on the right, found herself the subject of media condescension when she said in an interview that facial recognition and algorithms can be racially biased. The large amount of research on the subject, however, supports her claims.[45] AI constructed with the bias of its creators or those who train it can exacerbate racism, sexism, classism, and other problems rather than help to solve them.

Dumb AI

It is unsurprising that a technologist like Zuckerberg would propose a technological fix to the problem now colloquially known as "fake news" or the pressing issue of online hate speech, but AI is not perfect on its own. Ultimately, AI provides some technological fixes to these very real social problems while also presenting some fairly serious new issues. Meanwhile, the myopic focus of tech leaders on computer-based solutions is indicative of the naïveté and arrogance that caused Facebook and others to fail to protect users in the first place. Manipulative uses and flawed construction of technology are partly what got us into this mess. Ethics were not encoded into social media, while bias and arrogance were.

Again, it will always be people who are behind political manipulation, whether online or offline. There are not yet armies of smart AI bots working to manipulate public opinion during contested elections. Will there be in the future? Perhaps. But it is important to note that even armies of ML political bots will not function on their own. They will still require human oversight to manipulate and deceive. We are not facing an online version of *The Terminator* here. Luminaries from the fields of computer science and AI, including Turing Award winner Ed Feigenbaum and Geoff Hinton, the "godfather of deep learning," have argued strongly against fears that "the singularity"—the unstoppable age of smart machines—is coming anytime soon. In a survey of American Association of Artificial Intelligence (AAAI) fellows, over 90 percent said that super-intelligence is "beyond the foreseeable horizon."[46] Most of these experts also agreed that, when and if super-smart computers do arrive, they will not be a threat to humanity.

Stanford researchers working to track the state of the art in AI suggest that our "machine overlords," at present, "still can't exhibit the common sense or the general intelligence of even a 5-year-old."[47] So how will these tools subvert human rule or, say, solve exceedingly human social problems like political polarization and a lack of critical thinking?

The title of a 2017 article from the *Wall Street Journal* puts it succinctly: "Without Humans, Artificial Intelligence Is Still Pretty Stupid."[48] Grady Booch, a leading expert on AI systems, is also skeptical about the rise of super-smart rogue machines, but for a different reason. In a TED Talk in 2016, he hit the nail on the head, saying that "to worry now about the rise of a superintelligence is in many ways a dangerous distraction because the rise of computing itself brings to us a number of human and societal issues to which we must now attend."[49] More importantly, Booch stressed, current AI systems can do all sorts of amazing things—from conversing with humans in natural language to recognizing objects—but these things are decided upon by humans and encoded with human values. They are not programmed but rather, more aptly, they are taught how to behave. Booch elaborated:

> *In scientific terms, this is what we call ground truth, and here's the important point: in producing these machines, we are therefore teaching them a sense of our values. To that end, I trust an artificial intelligence the same, if not more, as a human who is well trained.*

So the picture of the relationship between machines and people becomes more nuanced. Crucially, Booch's argument further cements the point that smart systems and the people who build them and interact with them are inextricable.

I agree with Booch, though I would take this idea of his even further. To address the problem of computational propaganda we need to zero in on the people behind the tools. Computational propaganda, despite its wonkish name, is more about propaganda than about computers. Yes, ever-evolving technology allows people to automate their ability to spread disinformation and trolling and to operate anonymously and, with virtual private networks (VPNs), without fear of discovery. But this suite of tools as a mode of political communication is ultimately focused on achieving the human desire of control. Propaganda is a human invention, and it's as old as society. This is why I have always focused my work on the people who make and build technology. As an expert on robotics once told me, we should not fear machines that are smart like humans so much as humans who are not smart about how they build machines.

What happens as AI improves and the whole range of political actors, from governments to hacking collectives, use it to amplify or streamline computationally enhanced propaganda? With AI, governments or militaries could gather and parse data on entire populations and then use granular findings on individual citizens to drive political marketing over social media. Or smart systems could use amalgamated information on news events in a country to generate hyperrealistic fake-disaster scenarios and target them at those most likely to believe them. It's time that we use both human labor and machine power to prevent such scenarios, which have not yet come to pass and do not need to.

From AI to Fake Video

The persuasive power of AI is beginning to move away from the written word. Many argue that we are leaving an era of computational propaganda defined by bots peddling false news on Twitter and moving toward a time when convincingly fake AI video shows our politicians doing awful things. "Deepfakes"—doctored videos that use AI technology to make it seem as if someone is doing or saying something they didn't do or say—are beginning to hit the public web. Humor videos showing politicians saying ridiculous things have already showed up and been widely shared on YouTube. More nefarious and manipulative AI-altered clips of people doing things they certainly did not do are also beginning to surface.

Many journalists and pundits have argued that the deepfake is the next big thing in disinformation. They say that when these videos become mainstream, they will further erode the truth. How, after all, can the public vet a video that shows a public figure breaking the law if they think any video could be fake? How are deepfakes being used to spread propaganda today? Will they really lead to new, even more pervasive and persuasive waves of junk news? How can we track these AI videos? How should we respond?

Chapter Five
Fake Video: Fake, But Not Yet Deep

Doctored Videos Versus Deepfakes

CNN reporter Jim Acosta was never President Trump's favorite White House correspondent. In May 2016, as a presidential candidate, Trump sarcastically replied to a question from the reporter with, "I've watched you on television. You're a real beauty."[1] In 2017 Trump told the journalist, "You're fake news," after he pushed back on the president's assertion that *BuzzFeed* was "a failing pile of garbage." But in 2018 something different happened. After yet another contentious press conference exchange between the commander in chief and the CNN reporter, in which Trump called Acosta "a rude, terrible, person," a White House intern tried to take the mic away from Acosta as he began to respond. Acosta didn't immediately hand it over.[2] Later the conspiracy website Infowars posted a video clip that many conservative pundits said showed Acosta assaulting the woman during the interaction.[3]

Standing behind the allegations of assault, the White House suspended Acosta's White House press credentials. Press Secretary Sarah Huckabee Sanders said that he was punished for "placing his hands on [the] young woman."[4] She used the video clip as proof, going so far as to share it on Twitter.[5] The problem, however, was that the video in question was doctored. A number of publications, riding the wave of paranoia around deepfakes—or AI-manipulated videos—claimed that this was a clear example of just such a hit job. Soon, however, it transpired that the video had simply been sped up to make Acosta's attempt to hold on to the mic look hostile. This clip was more a shallow fake than a deep one.

Trump adviser Kellyanne Conway, the creator of the term "alternative facts," attempted, and failed, to suggest that the clip was not changed. "That's not altered. That's sped up. They do it all the time in sports," she told *Fox News Sunday* host Chris Wallace.[6] Of course, a video that is sped up is, in fact, altered. The White House was forced to relent, though they had done a thorough job of impugning Acosta's character. The CNN journalist's press passes were reinstated, and after a couple of weeks the media storm around the incident died down. For the pundits, however, there was one important takeaway from the incident: video was the next frontier for disinformation.

The Deepfake

Imagine this: It's presidential primary season, and all of the candidates are taking every chance they can to appear on TV, to host rallies, and to spread social media messages containing their campaign platforms. A video of the front-running Democratic candidate surfaces online and proceeds to go viral. You can't go on Facebook, Twitter, or YouTube without finding a link or reference to it. The video shows the candidate talking to a couple of people at a cocktail party. With her face looking directly into the camera, she clearly says, "I'd do anything to win an election, and I'm certainly not above taking a bribe."

There is an old joke that goes something like this: "How do you know if a politician is lying?" The answer: "Their lips are moving." True or false, funny or trite, there is a frightening nugget of foresight in this wisecrack. What if we could use technology to actually manipulate someone's lip movements in a video? What if we could do it with a family member or a friend? Or someone in a position of power? What if this same technology could also be used to change body movements? Or the background of the video? These are some of the questions behind the global concern over the deepfake.

In the past few years of conversation about the rise of digital disinformation, deepfakes have taken center stage. A portmanteau of "deep learning" and "fake," these AI-doctored videos present new, multi-sensory ways of manipulating perceptions of reality. Many researchers and experts suggest that they will make it nearly impossible for us to know whether a politician—or anyone else who has been filmed, for that matter—actually said or did what is depicted in a video. Sam Gregory, program director of Witness, a nonprofit organization focused on videos and human rights, describes the perceived threat of the deepfake:

> *Scenarios for the usage of synthetic media (including the ability to plausibly manipulate facial expressions in video, synthesize someone's voice, or make subtle removal edits to a video) include "floods of falsehood" created via computational propaganda and individualized microtargeting, which could target discrete individuals with fake audio and overwhelm fact-finding and verification through sheer volume of manipulated content.*[7]

Gregory alludes here to the potential political uses of this technology harnessed to a set of more conventional digital manipulation tools, including microtargeting and large volumes of computational propaganda. The use of deepfakes and other high-quality fake video, as with these other technologies, has serious ramifications for politics, but also for trust and security in business, civil society, the arts, and—more generally—everyday life.

Deepfakes could be used to challenge public opinion and what we know as reality in basically all sectors of culture and society. The question is, are deepfakes actually here? To better understand the disinformative threat of the deepfake—and determine if there actually is cause to worry—we have to ask ourselves a few supplemental questions: Where did these bogus videos come from? How do they work? Have they actually been used to manipulate public opinion or political conversations?

The unseemly truth is that deepfakes have a history of use for the creation of fake porn videos. In the simplest cases, a realistic version of a person's face is superimposed on another person's body. At present, celebrities are the most common victims of these face-swaps. The first examples of the term "deepfakes" being used, along with more sophisticated doctored pornographic videos, emerged on Reddit in 2017.[8] An account on that platform with the user name "deepfakes" uploaded its first manipulated videos, which were created using Google's "Open Source TensorFlow" AI tool.[9] The anonymous user posted their methodology, alongside the code they wrote using the software, to the same social media site.[10]

Before it showed up on Reddit, the relatively new technology was mostly consigned to the artificial intelligence research community. Now thousands of deepfake porn videos are available online. The *Guardian* succinctly describes the technical and historical origins of the tool:

> *Fake videos can now be created using a machine learning technique called a "generative adversarial network," or a GAN. A graduate student [and now research scientist at Google], Ian Goodfellow, invented GANs in 2014 as a way to algorithmically generate new types of data out of existing data sets. For instance, a GAN can look at thousands of photos of Barack Obama, and then produce a new photo that approximates those photos without being an exact copy of any one of them, as if it has come up with an entirely new portrait of the former president not yet taken. GANs might also be used to generate new audio from existing audio, or new text from existing text—it is a multi-use technology.*[11]

This technology is still mostly used in porn (despite moves by sites like Pornhub, Twitter, and Reddit to ban the abusive fake videos on their platforms).[12] There are few examples of deepfake technology actually being used for political manipulation. But those few have been potent.

The first comes from an unlikely place—comedy. In April 2018, Oscar-winning filmmaker Jordan Peele, working with *BuzzFeed*, used a combination of AI and non-AI tools to create a fake public service announcement (PSA) from former president Barack Obama.[13] In the now-viral video, Obama convincingly appears to say, "President Trump is a total and complete dipshit." The manipulated Obama also speaks to the fact that we now live in an era when our enemies can make us say or do anything. Peele used the video as an opportunity to warn the public about this burgeoning technology. The video was openly presented as a deepfake when released, but what if it hadn't been? What if it had been picked up and presented by news outlets as a true statement by the former commander in chief? Come to think of it, how do we know a video like this won't be stripped of its context and turned into evidence of exactly that?

In 2018 a diverse group of researchers from places like Germany's Max Planck Institute, Technicolor, the University of Washington, and Stanford University tested the use of technology they called "deep video portraits" on several politicians, including Obama, Ronald Reagan, and Vladimir Putin.[14] Again, in this circumstance, imagery of the leaders was effectively doctored. Though the researchers—whose efforts were funded in part by Google and the tech company Nvidia—identified the potential for misuse of this new tool, they said that they remained optimistic about its use for creative cinematography and technological purposes. In an earlier project, two University of Washington researchers who contributed to the deep video portrait paper were more upfront with their concerns about this work: "You can't just take anyone's voice and turn it into an Obama video," they said. "We very consciously decided against going down the path of putting other people's words into someone's mouth. We're simply taking real words that someone spoke and turning them into realistic video of that individual."[15] Not everyone, however, has been so conscientious.

In May 2018, a video surfaced of President Trump in which he calls on Belgium to join the United States in leaving the Paris climate agreement.[16] At face value, this statement didn't seem that far off from something the president might actually push for. His government did, after all, withdraw from the agreement, and the president is well-known for suggesting that his methods and choices are "the best." Subtlety of message, however, is often a very dangerous tool for manipulating reality. It turned out that this particular clip of Mr. Trump was actually fake. He never said those words in that setting, or anywhere else.

The video was created by the left-wing Belgian political party sp.a, which allegedly made it as a critique, but not everyone caught the nuance. Technologically doctored videos like this set a dangerous precedent for using deepfake technology. These bogus videos put words into the mouths of politicians and others in attempts to sow discord, sway voters, and disrupt the news process. They threaten not just democracy but also our perception of what is real by undermining what we can accept as evidence and challenging what we believe to be true. Deepfake technology is a potent tool in the wrong hands.

It is important to note that the video of Trump and Peele's video of Obama are not without their bugs. If you look at the videos closely, you would probably guess that something is off—though you might not realize that they are completely fake. They're blurry and the speech patterns are strange, qualities that some might chalk up to poor buffering or bad audiovisual syncing. But these problems actually stem from the issues with deepfake technology. They point to the fact that AI video doctoring is still in the early stages of development. The fact that the video from Peele and *BuzzFeed* is slightly better than the Trump one suggests that the tool can be more effective if the person wielding it has solid know-how.

Don't Stress Just Yet

Tim Hwang, director of the Harvard-MIT Ethics and Governance of AI Initiative and former policy lead for Google's machine learning portfolio, is less worried about deepfakes than he is about the people who fall for them.[17] He points out that "purveyors of disinformation were manipulating video and audio long before the arrival of deepfakes; AI is just one new implement in a well-stocked toolbox." Hwang uses the sped-up Acosta video as an example of less technically sophisticated video propaganda. He notes that research and history show us that propagandists are pragmatic in their approaches to manipulating public opinion: they use the simplest and most cost-effective means in efforts to change hearts and minds. As I pointed out earlier, dumb bots are a good example of this pragmatism. They achieve the ends of their creators through sheer numbers and are cheap and easy to use.

Photoshop or GarageBand, say, can be useful means for such ends—and they don't require deep technical knowledge or expensive training to use. Both deepfakes and the underlying machine learning and computational power needed to power them, like AI-driven political bots, are comparatively expensive and technically complex. As Hwang puts it, "Machine learning is a powerful but ultimately narrow tool: It creates better puppets, but not necessarily better puppeteers. The technology is still far from being able to generate compelling, believable narratives and credible contextual information on its own." The truth of this observation is readily apparent when comparing the video from sp.a and the one from Peele and his collaborators—people still need resources to produce convincing propaganda whether or not they use video.

But like most of the digital tools that have come before it, as deepfake technology progresses it is likely to become cheaper and easier to use. It is also likely that more sophisticated versions will become available. Even now, there are smartphone apps like Xpression that allow regular people to create lo-fi versions of doctored videos. Moreover, it's possible that people will reuse and remix more convincing deepfakes. What would happen, for instance, if someone took Peele's video of Obama, shortened it, and posted it to YouTube? Might that not enrage a community that was unfamiliar with the original video? In similar circumstances, repurposed propaganda has even led to offline violence.

For Hwang, the underlying psychological and sociological issues that lead people to believe in things like deepfakes, political bot campaigns, and other forms of computational propaganda are of more concern than the technology itself. The development of tools to detect digital disinformation, he points out, is progressing simultaneously to development of the tools for peddling it. We must, in other words, give as much attention to the demand side of propaganda as we do to the supply side. Who consumes junk news? What makes certain groups of people more likely to buy in to a conspiracy theory or a false news report? As I pointed out earlier in the book, critical thinking and conspiracy thinking are two sides of the same coin, though markedly and dangerously different. Both make use of evaluation and critique to interrogate things beyond face value. Conspiracy theorists, however, often substitute conjecture and rumor for scientific knowledge. Nevertheless, many conspiracy theories are widely popular, and some such rumors even go viral. The debunked "chemtrails" conspiracy, which asserts that airplanes' condensation trails are actually chemical agents being spread by governments or corporations for nefarious purposes, is still widely shared online today.

Hwang is right to question whether deepfakes will ultimately gain potency as propaganda tools. But I would add two key qualifications. First, well-resourced groups with a stake in politics—from militaries to corporations—certainly have the technical and financial resources to leverage sophisticated forms of disinformation. While it's true that in 2016 Russia mostly used clunky social media bots to manufacture trends on Twitter or fake grassroots group pages to sow political infighting on Facebook, it is also true that the era of such basic digital methods is ending. Yes, these companies and a slew of other groups are working tirelessly to create tools to stop these particular misuses of social media and the internet, and as they do so, the companies and their technical infrastructure are evolving. History, however, as well as information operations doctrine, suggests that those working to manipulate through the media are often one step ahead of those working to protect against such manipulation. As social media platforms change, so too will the ways in which computational propaganda is spread. The technology that is costly today is likely to be cheap and easy to use tomorrow. Moreover, disinformation campaigns are nearly impossible to put back in the box once they have been launched. Research shows that post-hoc efforts to reeducate or fact-check online misinformation or disinformation often fall flat.[18] Researchers Brendan Nyhan and Jason Reifler have argued for the prominence of "a 'backfire effect' in which corrections actually increase misperceptions among the group targeted."[19]

Beyond this, deepfake technology is already being used by the powerful to spread disinformation. They just might not be the same powerful actors—or even the same uses—that AI experts first had in mind when thinking through potentially problematic uses of AI video technology. The use of deepfakes in and around pornography exposes a seriously problematic and power-laden use of technology in which one person releases a fake video that subjugates, harasses, and shames another. One video that appeared on Reddit, made with a program called FakeApp, combines both political and pornographic disinformation. In it, former First Lady Michelle Obama appears to do a striptease. According to the *New York Times*, "the hybrid was uncanny—if you didn't know better, you might have thought it was really her."[20] We must accept that these uses of deepfakes are as problematic as those that show Donald Trump or some other male politician saying something stupid or irreverent, and perhaps more so. Both uses of deepfake technology must be stopped before they take hold.

Most often, it is the victims themselves who are left with the nearly impossible task of getting the video removed from the internet. Not only does this show just how resilient disinformation can be, but it begins to reveal the psychological stress that victims of computational propaganda must carry. Although future deepfakes might not change the outcome of elections, they will undoubtedly have other repercussions for victims, no matter who they are. If we don't build safeguards into new technologies before we release them into the wild, then they will undoubtedly be misused. There are ways we can use code-level analysis to track deepfakes and other technology used to spread propaganda, and there are even ways to track these AI-manipulated videos using manual qualitative searches. Social media companies and other groups must invest in this technology now. Those experimenting with and building new AI video technology must think very carefully before they release their creations to the public. We need a shift in how companies build their products and what consumers expect of them. It is not enough to presume that, just because polarization and disinformation are born of social issues, technology isn't adding to the problem.

Regular Videos: Another Powerful Propaganda Tool

AI video will undoubtedly continue to become a potent tool for spreading propaganda, but even real videos are useful to this end. Clips that are selectively edited and spliced together, as happened with the Acosta reel, can highlight a gaffe or take something out of context. Anything that appears to be evidence can be a dangerous tool for those hoping to alter the truth or rewrite history.

BuzzFeed opinion editor Tom Gara makes the argument that, at present, we should be more concerned about real videos than fake ones.[21] According to a report on the state of online video from Limelight Networks, a content delivery network (CDN) company, from 2017 to 2018 the average global user increased the time they spent watching video each week from one hour to six and a half hours.[22] Limelight found that "people aged 18–25 watch an average of almost nine and a quarter hours weekly, with 31 percent of them watching 10 or more hours." Cisco's Visual Networking Index report forecasts that online video traffic will grow fourfold from 2017 to 2022.[23] Video is popular, and it only looks to get more so, with platforms like TikTok on the rise.

The ubiquity of the medium aside, Gara points to the highly publicized 2019 incident in Washington, DC, between high school students and Native American elders as an example of the misinformative potency of regular online video. The situation in question arose after the release of a video clip that appeared to show a group of students from Covington Catholic High School in Kentucky who were wearing MAGA hats and mocking a Native American man during a protest at the Lincoln Memorial. Subsequent clips of the alleged confrontation revealed a more complex story. Some argued that these clips showed that the students were peacefully if somewhat ostentatiously chanting their school song. Others said that the clips showed that a group of leftist protesters were actually the ones inciting violence.

Gara points out that there were neither AI alterations to the video nor any basic edits beyond showing particular clips of a longer scene. In fact, he argues, "throwing video into the mix can make the picture even murkier, giving everyone the opportunity to find the split second that makes their case; the new angle that raises new questions."[24] He continues: "There might be nothing as misleading as an extremely genuine fragment of extremely genuine video."

Another example of an edited video being used for divisive political means arose in mid-2018. *Conservative Review TV* (CRTV) created a doctored clip of Alexandria Ocasio-Cortez in which the then-congressional candidate appeared to be at a loss for words during an interview. According to the *Intercept*, "the fictional interview with one of the network's hosts, Allie Stuckey, was created by splicing in answers Ocasio-Cortez had given to different questions during a recent appearance on the *PBS* show 'Firing Line.'"[25] The problem was that when CRTV initially posted the fake "interview" to social media, it did not alert viewers to the fact that it was manipulated. The clip quickly amassed over one million views. CRTV later claimed that it had made the video as a satire rather than as a source of disinformation. Sound familiar? In a tweet, Ocasio-Cortez responded to the video saying, "Republicans are so scared of me that they're faking videos and presenting them as real on Facebook because they can't deal with reality anymore."[26]

When basically edited videos, like the clips of the Covington Catholic students and Ocasio-Cortez, are put on a site like YouTube, they have as much potential to go viral as complete videos, if not more. What mechanisms are in place to make sure that doctored or out-of-context videos do not get posted? Are there checks and balances? Why is video particularly convincing?

The multi-sensory nature of video makes it a strong tool for spinning the truth one way or another. According to research, video is a more potent tool than text for spreading ideas because it more effectively stays in one's memory. Video apparently has an advantage over simple words in this regard because you see *and* hear the content. It is made that much more real by our ability to witness events with our eyes and ears. According to a guide from the corporation 3M on how to give effective presentations, the brain processes video 60,000 times faster than it does text.[27] What is more, research from two Australian scholars suggests that motion can be a very effective means of capturing and maintaining attention.[28] It's no surprise, with this in mind, that TV shows and movies now constantly cut from one thing to another. It's also not shocking that propagandists have picked up on video as a tool for spreading their bespoke versions of reality.

Video has something else going for it as an effective propaganda tool: it often uses captivating modes of storytelling to get a point across. In a seminal paper on the subject, researchers showed that psychological transportation, defined as "absorption into a story," is a useful tool for persuasion when people read public narrative content.[29] The authors of the study explain that transportation "entails imagery, affect, and attentional focus." They point out that "highly transported readers found fewer false notes in a story than less-transported readers." Video, with its attention-grabbing features, has proven to be especially powerful at transporting those who watch it.

YouTube faces major hurdles in vetting the content that gets put on its site—as do platforms including Instagram, LinkedIn, Video, TikTok, Snapchat, Vimeo, and WeChat, all of which now prioritize the creation and use of social video. Many video websites do employ both people and algorithms to vet and monitor what is posted. But most people wouldn't have caught and tagged the Covington Catholic video as problematic. Nor would algorithms have caught this politically sensitive video. Should people or code have stopped it from being aired in a fragmented clip? This is a challenging question because such restriction is very much at odds with the free speech via video that YouTube, in particular, claims to prioritize. This said, the video platform can and must take clear steps to prevent the posting of more videos that have clearly been doctored. And platforms should think long and hard before allowing deepfakes of any kind to be posted. The question is, can monitoring teams identify these videos effectively enough to stop them?

The YouTube Problem

YouTube largely escapes notice in many of the computational propaganda events I've sat in on. Twitter, Facebook, and even Reddit are more readily offered up as examples of platforms where computational propaganda and other forms of technically enhanced manipulation run rampant. But YouTube has its fair share of problems, most notably its history of placing disinformative and conspiracy theory–laden videos at the top of trends on its home pages and quickly linking people to extremist videos after they've watched less partisan ones.[30] Some of these videos are also amplified by bots that link and spread them across the internet and other social media platforms, giving them further currency among the general public.

Researcher Becca Lewis has demonstrated that YouTube is a favorite medium of white nationalists and the alt-right. In her report "Alternative Influence: Broadcasting the Reactionary Right on YouTube," Lewis argues that the website's facilitation of the platforms for numerous extremist "influencers" amounts to what she calls an "alternative influence network."[31] She writes that, "by connecting to and interacting with one another through YouTube videos, influencers with mainstream audiences lend their credibility to openly white nationalist and other extremist content creators." She also points out that all influencers, including those who peddle far-right conspiracy theories, can and do use YouTube to make money—sometimes in very large amounts.

But the problems with YouTube are not only associated with its use to spread polemic and, at times, disinformative content from both the right and the left. Nor are its problems tied to the fact that people can make money by doing so. There are other issues similarly grounded in the way YouTube functions. These problems exist both in the code and in the human-based guidelines for prioritizing certain content over others.

According to the Google News Initiative, YouTube does in fact organize breaking news and other content based on quality metrics, including perceptions of the authority of the news outlet.[32]

It has actually begun to take credit for doing so as part of a multimillion-dollar endeavor aimed at building bridges between YouTube and the news media. In fact, *Mashable* reports, the media giant has now started showing breaking news at the top of its search mechanism.[33] At first glance, the company is working to prioritize content that it deems newsworthy or "authentic" instead of sharing conspiracy theories or disinformation. But underneath this appearance, the company is still quietly curating the news content that people get to see. Perhaps YouTube should let us all in on how it makes these choices?

In some ways YouTube is damned if it does and damned if it doesn't. If it leaves manipulative or misleading content on its site, it is criticized. If it curates such content, it is similarly attacked. YouTube is clearly taking action, but some elements of transparency are missing from its efforts. There is a lack of clarity about *how* YouTube works to rank videos. What role do algorithms play in this practice? What is the role of human moderators? What signals are used for ranking news content? Are they all Google-based, or does the firm work with other social media companies?

Beyond this, the company still treats both US and global problematic content that it identifies as "political" but not disinformative in a piecemeal and opaque fashion. It has been established that companies, including Google, embed themselves with political campaigns in order to consult and sell ad space.[34] How then can we trust them to effectively self-regulate or moderate political content?

Google has recently instituted the use of "Knowledge Graphs"—information panels that appear on the side of search results for key historical events and other topics. If you search "the Cold War," you will see images related to the topic, as well as key dates and a link to a description of the event from Wikipedia. Perhaps it is time for YouTube to take similar measures in regard to "key event" or "scientific topic" searches on its platform. In early 2018, YouTube began a new policy of linking to Wikipedia articles as a means of combating conspiracy videos.[35] The problem was that the social media company didn't let the not-for-profit encyclopedia know about the move. According to the *New York Times*, the move by YouTube has threatened to burden the "humble" information-oriented website.[36] The newspaper summarizes the problem perfectly: "Here was Google, a company with revenue in excess of $100 billion last year, calling on a volunteer-built, donation-funded nonprofit organization to help it solve a crisis." In response, Katherine Maher, executive director of Wikipedia's parent organization, the Wikimedia Foundation, said, "When the announcement came out, we were surprised that we hadn't been contacted."

If Google and YouTube are going to use Wikipedia as a mechanism for showing trustworthy content on certain topics, then they should be generously compensating the organization, with no strings attached. Although Wikipedia is one of the most accessed sites on the internet, it is supported by its users and by a relatively small group of very active article writers, or "Wikipedians." If tech giants continue to look to not-for-profit or news organizations to combat disinformation and check facts, then they must include and support these organizations fully. Facebook and Google must trust organizations such as Snopes, the Associated Press and other journalistic entities with the full purview of data on a given subject.[37] They must generously fund such initiatives with no strings attached—no editorial oversight of the outside organizations' content or day-to-day operations. Otherwise, we run the risk that tech firms will exert unfair control over and extract unpaid labor from these smaller entities.

As I mentioned earlier, there are people who rate and moderate videos on YouTube. Who are these people? Are they Google employees, or are they contractors? If they are contractors, do they receive the same benefits as Google employees? How are their biases mitigated as they moderate? Both research and statements from former moderators have established that they are often poorly paid and forced to curate disturbing content.[38] YouTube must release more information on who these moderators are, demographically, geographically, and otherwise. The social media platform, and others like it, must also ensure that bias among its moderators is mitigated and that they use a quality bar for rating content. These platforms should publicly release information on intercoder reliability among moderators—that is, whether those coding or ranking data consistently agree or disagree about the quality of data. Perhaps most important, the platforms must protect these moderators and compensate them fairly.

It's also important for us English speakers to remember that YouTube and other social media platforms provide content in a wide variety of languages. The experts I've interviewed have told me that, until very recently, most of the companies were struggling to keep up with the problem of false information in English alone. How, then, could they be effectively preventing the flow of propaganda in hundreds of other languages? What about languages not yet translated into computer code? Recent reports suggest that the firms are scaling their attempts to address the problem, but with such massive user bases, how many moderators and researchers will be enough? This language problem also comes into play in efforts by Facebook and other social media firms to consider cultural context when preventing the spread of malicious or politically manipulative content. It is challenging enough to stop the flow of rumors if you do not understand the culture from which they are coming. What if you speak the language but are not a local? In this scenario, you probably wouldn't understand the intricacies of the culture. Progress in textual analysis and natural language processing has now made it possible to more accurately translate written text. With video, however, these efforts are even more challenging, though not impossible.

Stopping the Spread of Fake Video

With video becoming so pervasive online, how can we stop its use for the purposes of propaganda? Are there tools that have been built to detect deepfakes? What about video that has been doctored in more rudimentary ways? And what about streaming video? Facebook Live, Instagram, and Periscope all offer easily accessible applications for live-streaming, and this technology is on the rise. How could live-streaming lend itself to manipulation, and how might we prevent that?

Like any technology or media, the internet will continue to grow larger and more sophisticated unless something drastic happens. It is more likely, for instance, that the hundreds of undersea cables that provide global connectivity to the web will consistently be replaced and improved upon than that they will suddenly be disconnected in some way. Short of nuclear war or a similarly apocalyptic event, the internet is here to stay. And online video is also here for the long run. But if it is possible to use video to spread propaganda and digital hate, it is also possible to institute measures to stop such misuses. Because deepfakes make use of AI-enabled algorithms, we can develop tools to help us unearth these doctored videos—by tracking their hallmarks.

Some groups are already working on algorithms made to detect, identify, and combat deepfakes. A group of researchers from the State University of New York at Albany's Computer Vision & Machine Learning Lab have built a program that detects deepfakes using a fairly unique method. The team monitors video subjects' blinking patterns.[39] This remarkable technology is built on the premise that if deep neural networks allow deepfake makers to falsely alter people's facial expressions, then there must be some sort of tell. According to the lead researcher on the project, Professor Siwei Lyu, "when a deepfake algorithm is trained on face images of a person, it's dependent on the photos that are available on the internet that can be used as training data."[40] What is more, he says, "even for people who are photographed often, few images are available online showing their eyes closed." Because of this, the team is able to catch deepfakes. As Lyu explains, "When we calculate the overall rate of blinking, and compare that with the natural range, we found that characters in deepfake videos blink a lot less frequently in comparison with real people."

And with this small discovery about the training mechanisms behind deep learning, those working to combat deepfakes have one more tool to catch these AI-doctored videos before they take hold. Many of the other methods for detecting these bogus videos look beyond blinks. Some of them are even human-based. There are several ways in which a deepfake can be built, or particular areas that the builder can focus upon. For instance, the builder could focus on only changing lip movement—though the result when using this method alone is usually less than convincing. The builder could also, as I mentioned earlier, simply swap the face of the person being captured. This building technique is similar to the relatively rudimentary tool provided by applications such as Snapchat.

The deepfake builder can also do what Professor Lyu looks out for: use multiple photos to alter facial expressions in more sophisticated ways. Finally, the builder can use the motions of one person to emulate the similar motions of another person.

All of these methods for constructing deepfakes, according to the *Wall Street Journal*'s Ethics & Standards team and its Research & Development team, have their own tells. The *Journal* has been training its staff to recognize deepfakes using particular markers. Beyond simply raising awareness among reporters and editors that the technology exists, the newspaper is using a combination of technological methods and old-school reporting techniques to suss out fake video. According to Natalia Osipova, a *WSJ* senior video journalist, "There are technical ways to check if the footage has been altered, such as going through it frame by frame in a video editing program to look for any unnatural shapes and added elements, or doing a reverse image search."[41] But the best option, the *WSJ* staff argue, is usually to simply "reach out to the source and the subject directly, and use your editorial judgment." Reporters and regular people alike should always examine a questionable video source. They should look for older versions of the video. Finally, they should examine the video closely in an attempt to see if there are fuzzy frames or other actions that are out of place. Again, it is a combination of human and technological strategies that can be brought to bear on this problem.

The deepfake detection company Amber Video is also working on novel ways of stopping the flow of these videos before they start. CEO Shamir Allibhai and his colleagues have created tools, including Amber Authenticate, that catch manipulation by logging the original video in the blockchain. According to *Wired*:

> *The tool is meant to run in the background on a device as it captures video. At regular, user-determined intervals, the platform generates "hashes"—cryptographically scrambled representations of the data—*

> *that then get indelibly recorded on a public blockchain. If you run that same snippet of video footage through the algorithm again, the hashes will be different if anything has changed in the file's audio or video data—tipping you off to possible manipulation.*[42]

Allibhai also argues that there is a need for a code-based mechanism to mark videos with stamps when they are first created. Conversely, there could be a way to mark all deepfakes when they are first created—but this would be likely to require buy-in from the hardware and software companies whose devices and programs are used to create these videos.

The Problem of Live-Streaming

Live-streaming is a growing medium of choice among the many who feel that it is more authentic than edited clips or images that one might post on social media. Could hackers or propagandists manipulate these real-time videos, however, to spread disinformation or political harassment? Could live-streaming give a platform to terrorists or violent groups hoping to gain public attention? In one fairly humorous case, a hacker known as Chang Chi Yuan, who gets paid to look for security flaws in websites, said that he would live-stream himself deleting Mark Zuckerberg's Facebook profile.[43] The bug hunter eventually backed out of the attempt, but might someone looking to engage in violence propose to do something similar?

In late 2018, a gamer on Fortnite did exactly this. The man in question, a twenty-six-year-old from Australia, live-streamed an attack on his pregnant partner—the mother of his two other children—while playing the game.[44] Other gamers reported the crime, and he was subsequently arrested. In another case, twenty-five-year-old Frenchman Larossi Abballa live-streamed portions of his murder of a French police captain and his partner.[45] The terrorism suspect was eventually shot and killed by the police, but the story got international attention. As *Forbes* points out,

"Abballa's video highlights the huge challenges technology companies, especially social networks, face monitoring and vetting live videos." In another case, the Russian government–run Russia Today (RT) network live-streamed protests in Rome that it alleged were against the Italian prime minister and a proposed referendum. But the protests were taking place for the opposite reason—to support the referendum. As *Wired*'s Bruce Sterling notes, "RT's live stream reached 1.5 million viewers and gave them the impression that the exact opposite was actually happening."[46]

These manipulations and attacks all happened via live-streaming before yet another horrendous, sociopolitically motivated killing spree: the Christchurch mosque attacks. The murderer live-streamed himself using Facebook Live during this mass shooting. Fifty people were killed and fifty were injured. Hundreds of people saw the stream and witnessed the hate-motivated killings live. After the attacks, Facebook said that it would consider restricting its live-streaming application. In a letter to the *New Zealand Herald*, Facebook's chief operating officer, Sheryl Sandberg, said that Facebook would explore "restrictions on who can go Live depending on factors such as prior Community Standard violations." She also noted that the video had been picked up by other users and reshared in over nine hundred other versions.[47]

The simple truth is that social media firms do not have to host live video streaming services. Nor, for that matter, must Facebook have a newsfeed that curates information.[48] Nevertheless, many of the major technology firms have already created some version of live-streaming, even though the task of moderating video as it is being filmed is very tricky. How might a company vet such content as it is being produced? Technology, including AI detection algorithms, can be useful in this regard because it can computationally enhance the rate at which problematic content is searched out. But technology will not catch all violent or misleading live-streams. It may also run the risk of falsely identifying many streams as violent or manipulative when they are not.

From Video to Virtual Reality

Video is already a well-entrenched and growing medium on the internet, as the statistics cited earlier suggest. New technologies and even novel media are just beginning to become widely connected. In the past few years, there has been a great deal of excitement about extended reality media: virtual reality (VR), augmented reality (AR), mixed reality (MR), and more. Enthusiasm about VR has recently abated to some extent because of problems with the price, mobility, and other aspects of the tech. But VR, AR, and MR are all in line to be the next big social media tools.

What would a VR or AR social media platform look like? Many of us will have heard of movies or books in which people all but live in VR worlds, logging off only to eat, use the toilet, and sleep. These platforms, until recently, have been consigned to the world of fiction. Experts argue, however, that the time when we'll interact with our social networks using extended reality media tools is drawing near. Not only has the technology advanced in recent years, but it is becoming more affordable. You need look no further than the latest gaming consoles to find VR in the average household. And a number of mobile phone and tablet games already include AR. You simply hold up your phone using the camera and behold digital creatures overlaid on the world around you. The wildly popular game Pokémon GO is a perfect example.

These tools present exciting new possibilities for socializing and entertainment. There are even people working to build interactive educational programs using VR and AR. But as communication tools, extended reality media can be used to manipulate as well as inform. The next chapter explores how VR and AR are being used to promote democratic values, to enhance gaming and movies, and to connect people. It also goes over the ways in which these media tools are already being used to spread computational propaganda. Finally, the chapter outlines ways in which we can preserve this emergent technology for positive societal uses while preventing large-scale misuse.

Chapter Six
Extended Reality Media

Virtual War

A group of soldiers walks through a battleground, each constantly looking from side to side to make sure enemies aren't approaching. They are outfitted with the latest in military gear and technology: advanced body armor, streamlined automatic weapons, and a variety of other tools and devices. Suddenly, a group of hideous humanlike creatures burst from hiding near the soldiers. They let out distorted cries and sprint every which way. It is unclear whether they are attempting to attack the soldiers or run away from them. The soldiers gun them down. Each and every creature is killed. But they are not monsters. They are regular people.

It turns out that the soldiers have been given neural implants called MASS, which use augmented reality technology to alter and enhance their senses and provide instantaneous data on their surroundings. Unfortunately, unbeknownst to the soldiers, the technology uses digital imaging tools to overlay enemy combatants—real people—with zombielike appearances and audio software to make them sound distorted and frightening. The implants use AR to dehumanize and deceive. The idea, from the commanders and governmental actors who oversee the military, is that soldiers will be more efficient killers if they see their enemy as creatures out of a nightmare. Later it is revealed that the soldiers are exterminating these people because of a eugenics program—an effort to create a homogenous master race. It is a worst-case scenario for the use of extended reality media tools.

Thankfully, this scenario is only that—a fictitious future in which extended reality tools are used in warfare to make for more moldable troops. It is from an episode of the popular television show *Black Mirror*, which often uses present-day problems at the intersection of technology and society to highlight potentially frightening, even dystopian, futures.[1] *Black Mirror*, and similar shows like *Mr. Robot* and *Humans*, are especially provocative because they tell stories about potential uses of existing technology. As we all know, AR is readily—if not widely—available in a variety of formats. Advanced humanoid robots, which feature prominently in *Humans*, are already being created by companies and university research teams.[2] The thing about the situations depicted on these shows in which current technology is used is that they hit very close to home. They could, with very little alteration, become reality.

As with any other communication tool, extended reality media can be used for less than inspiring causes. In fact, as with social bots or video, virtual reality and augmented reality can be exploited for coercion and influence campaigns. Virtual reality and the advanced technology that underpins it can be used to more effectively transmit computational propaganda. In the hands of people hoping to manipulate public opinion, these tools can provide multi-sensory disinformation or spin to receptive users.

Others around the world are beginning to realize the educational and even democratic power of extended reality media. The tools are particularly powerful in this regard because they put users in a digitally enhanced landscape where they can see, hear, and feel the elements of novel adventures. Many groups are beginning to use the technologies to teach people about historic events or to show them what it feels like to experience racism, sexism, or other forms of prejudice. Others are deploying them to reveal the perils of, for instance, environmental degradation and climate change. The idea is that if you put people in the midst of these experiences, they are better able to empathize with the plight of other people or far-off places. The science detailing these psychological effects is still progressing, but initial results suggest they are potent.

At present, there are more examples of positive and democratic uses of extended reality media than negative and manipulative ones. It should be noted, however, that most internet technologies have started out with a great deal of fanfare and hope for their perceived social benefits. Remember when the internet was going to be *the* tool to save democracy by making all information free and accessible? As with these other technological tools, negative uses of extended reality are on the rise.

The World of Extended Reality Media

The Institute for the Future (IFTF) has been doing forecasting work at the intersection of technology and society for over fifty years. Located in a large office space in the middle of Silicon Valley, in downtown Palo Alto, IFTF is home to a heterogenous group of anthropologists, game designers, artists, historians, journalists, and, unsurprisingly, futurists. The research institute split off from the Rand Corporation, a large global policy think tank, decades ago to do public-facing work on the National Science Foundation Network (NSFNET) and Advanced Research Projects Agency Network (ARPANET), precursors to the modern internet. Since that time, IFTF has collaborated on unique forecasting projects with a diverse range of people and groups, from corporations like 3M and Toyota to tech firms like Google and Mozilla, to philanthropies like the Lumina Foundation and the Bill and Melinda Gates Foundation.

A lot of interesting people pass through the doors of "the Institute." Jane McGonigal, game designer and author of the *New York Times* best-seller *Super Better*, is a research director there. Mark Frauenfelder, founding editor in chief of *Make* magazine and *Wired*, and David Pescovitz, journalist and founder of Omza Records, are also research directors. The two are also co-editors of the popular and pioneering blog *Boing Boing*. Marina Gorbis and Bob Johansen, executive director and distinguished fellow, respectively, have done innovative work on futures thinking with leaders of many of the world's largest companies, prominent civil society groups, and governments. Several labs do cutting-edge, futures-oriented research on different subjects: the Work and Learn Lab, the Food Futures Lab, the Emerging Media Lab, the Government Futures Lab, and the Health Futures Lab among them. And as of 2017, there is the Digital Intelligence Lab—of which I am the founding director.

Our work in the DigIntel Lab inevitably overlaps with some of the Institute's other areas of study. For instance, I've learned a lot from Toshi Anders Hoo, director of IFTF's Emerging Media Lab. Toshi was particularly instrumental in getting me to think about how immersive reality tools from VR to AR could be used for both societal good and political ill.[3] He was the one who told me that "the body has no metric for fake." He pointed out that while you might be able to see signs of manipulation in some media, like altered blinking patterns or strange lip movements in a deepfake, or hear them in others, like improper modulation of an automated voice, it is much more difficult to catch these tells when these two senses are combined with movement in an immersive media environment. Toshi and I have talked a lot about the future of the extended reality media space and how these tools could be used both for good and for bad. We've asked each other, as well as our colleagues at IFTF, a lot of questions: How could these tools be used to spread disinformation? How could they be used to further bias? How could they challenge notions of reality and the truth? On the other side of things, how could they be used to educate? How could they be useful for mental health therapy?

Defining the Virtual

Let's define terms before we get into the good, bad, and ugly of extended reality media tools. There is a lot of debate and confusion surrounding these terms, so let's agree to use these definitions for the duration of this book:

Extended reality, or **cross reality (XR)**, functions as a kind of umbrella term for all extended reality media tools that combine virtual reality with the real world.

Virtual reality (VR) is perhaps the best known of all extended reality media. Although often similarly used as a catchall term in the space, VR is actually specific to firsthand immersion in real-time, computer-generated, multi-sensory experiences. Think goggles and headsets, complete with earbuds, transporting the user to another place.

Augmented reality (AR) places digitally created imagery on top of the real world. A solid example of AR is the game Pokémon GO, which allows users to find capturable creatures in real-world spaces using the screen and camera of their smartphones.

Mixed reality (MR), another broad term, is commonly used in advertising in this space. Most accurately, MR is a hybrid tool that combines multiple types of extended reality media. It refers to media experiences with both VR and AR components. A set of glasses that blends both the real and virtual world would be considered MR. MR, in other words, allows the real and virtual to interact simultaneously.

Augmented or Virtual Reality?

A quick glance at reporting on these topics reveals a slew of articles claiming that VR is out and AR is in, conflating one with the other interchangeably or hyping the hottest new product in the space. Here my goal is to discuss how these somewhat interchangeable forms of media are currently being used not only to promote equality and educate people but also to exert power and control over people. Using current examples, I build scenarios in this chapter about the problems that could be posed by particular uses of these tools down the road. I also provide some concrete ideas on what we can do to prioritize efforts to mitigate political misuses of the tools while promoting their use for democratic purposes.

The extended reality media market is somewhat in flux. In 2016 and 2017, you almost couldn't open a magazine or newspaper without seeing an article about virtual reality or augmented reality. Recently, though, several factors have leveled out the excitement in the space. According to Andy Kangpan, an investor at the firm Two Sigma Ventures, VR in particular is in the midst of a "public relations slump."[4] Besides the waning enthusiasm around VR, he points to several key problems in the space:

> *The all-in cost for state of the art headsets is still out of reach for the mass market. Most "high-quality" virtual reality experiences still require users to be tethered to their desktops... When it comes down to it, the holistic VR experience is a non-starter for most people. We are effectively in what Gartner refers to as the "trough of disillusionment."*

But while it may take longer than people previously thought to get high-quality, cheap, mobile, and easy-to-use headsets to market, VR and the wider tools of immersive reality still look to be major players in the future media space.

Kevin Kelly, author and founding executive editor at *Wired*, argues that it is AR and not VR that will be the next big media technology.[5] He imagines a future in which we would all be able to experience digital re-creations of the real world, simulated twins of the physical. He describes this place as "mirrorworld," borrowing the term used to refer to this particular brand of digitally enhanced reality coined in the early 1990s by Yale computer science professor David Gelernter.[6]

Kelly's belief that the next big technology platform will feature AR encounters some disagreement among Silicon Valley leaders. Investor and software engineer Marc Andreessen has argued that VR will be a thousand times bigger than AR in the long run.[7] He points to the fact that VR, unlike AR, is fully immersive. AR may spark short-term interest, but it is

VR that will be more attention-grabbing. And while it is likely that VR will ultimately be the major innovation in the XR space, it is probable that all tools will eventually operate from the same space—as three different ways to access the metaverse. Each XR tool is here to stay.

The use of immersive reality already extends well beyond the gaming sphere. Artists, educators, activists, and others are making use of these tools for a variety of other purposes. In 2017, filmmaker Alejandro González Iñárritu's *Carne y Arena* (*Flesh and Sand*) was the first VR project to be accepted into the Cannes Film Festival, and it was awarded the first special achievement Academy Award in twenty years. The VR film was based on the director's interviews with Central American and Mexican refugees. It places the viewer amid a group of people as they attempt to cross into the United States. Similarly, several organizations and technologists are working to create extended reality media products that educate people about difference and a panoply of diverse experiences. Others are building tools that help people learn about other cultures or systems of belief.

But if VR and AR and other tools like them can be used for good, they can also be used for other purposes. What happens to truth—or to reality—when people use these tools to manipulate public opinion? If young people are immersed in multi-sensory experiences that teach, say, alternative histories that prioritize white supremacy, how will this affect their view of the world? Will people begin to build extended reality media tools that work to indoctrinate people to their point of view? What would these machinations look like? We may find that the possibilities are not a far cry from the types of campaigns we already see today on social media.

Extended Reality Media and Manipulation

It is not a foregone conclusion that virtual reality and augmented reality will continue to be used primarily as entertainment technologies, let alone that they will only be harnessed for the good of society. Bias can be built into the very code that underlies extended reality media. Hardware tools, from goggles to haptic suits, can be built to favor one body type over another. Discrimination and trickery can also be built into the multi-sensory XR experiences available to a user. VR and AR have already been used as tools for political manipulation. Without ethically focused design, and absent informed policy around particular uses, this technology could be the future of indoctrination and deceit.

In China, the ruling Communist Party has already used VR experiences to test and cement political loyalty.[8] In Shandong province in eastern China, members of the party have been required to submit to loyalty tests using VR. According to the popular industry blog *VRFocus*, "the 'Test of Dangxing' or Test of Party Spirit, was taken by party members in Qingyang, who were required to wear VR headsets and enter a virtual room, where they were quizzed on a variety of subjects, including party theory, members' daily lives and how they understood the 'pioneering role' of the party."

Making party members appear in a virtual room might, at first, seem to be overkill. Why conduct a "loyalty test" this way rather than via a phone call or video conference? Answers to this question may be inconclusive, but an obvious one is that VR is more immersive, and therefore more threatening. There is no computer scrolling option, as you might have on a conference call. Immersive testing also allows party leaders to track the test results by proxy, without having to travel to the test-taker's location. The consequences of failing the test, or behaving dishonestly, presumably feel more real when you are in a multi-sensory environment. According to the group that administered the tests, the results will be used to inform and identify the people and particular qualities that the party wants to promote. VR, in short, is a superior medium for manipulating others.

Could similar extended reality media experiences be used to spread racism or partisan politics in democratic countries? Palmer Luckey, creator of Oculus VR and designer of Oculus Rift—and one of America's richest individuals under the age of forty—is known to have funded at least one dubious social media propaganda campaign. The *Daily Beast* reported that the Silicon Valley wunderkind covertly provided thousands of dollars of financial support to Nimble America, a "social welfare non-profit", during the 2016 US election. The group allegedly had an initial focus on creating anti-Hillary Clinton billboards. According to the *Daily Beast*, however, the group intended to focus on the power of memes and shitposts (poor-quality online trolling) related to the contest.[9] The group was founded by two moderators of the Trump-supporting subreddit r/The_Donald, a well-known breeding group for just this type of content. Luckey told the *Daily Beast* that he used an online pseudonym, "nimblerichman," provided by the company for use on Reddit.

In 2017 the *New York Times* reported that Luckey had bounced back from the industry isolation imposed by largely anti-Trump sentiment in the technology sector (Luckey also provided $100,000 for Trump's inauguration) to begin a new start-up, this one focused on a "virtual border wall."[10] The company, funded in part by conservative tech billionaire Peter Thiel, was to focus on building security technology to monitor the country's borders and military bases. According to the *Times*, Luckey's new venture "plans to use a technology found in self-driving cars called lidar—shorthand for light detection and ranging—as well as infrared sensors and cameras to monitor borders for illegal crossings." In 2019 Luckey told CNBC that the company, under the name Anduril Industries, had already begun testing its border control technology—Lattice—in California and Texas.[11]

What is to stop a near-billionaire like Luckey—or the Facebook board member, Trump technology adviser, and Cambridge Analytica funder Thiel—from using VR and AR technology to similarly spread propaganda? Should either these individuals or those on the far left of the political spectrum be allowed to use extended reality tools for such purposes? Are these uses protected free speech? Do they violate privacy or human rights? If VR and AR continue to be used manipulatively, will XR be the next technology to "break" democracy and the truth?

These questions may make for interesting philosophical debate, but what's crucial right now is that they have not been meaningfully answered by our lawmakers or the technology firms creating the products. Until regulators begin to generate systematic and informed laws about extended reality media technology, it will be up to Silicon Valley and its leaders, including Luckey and Thiel, to dictate not only how technology is used but how it will be developed. What tools will these companies focus on building? What kinds of experiences will they prioritize? Which employees will be tasked with creating the hardware and software that define them? At the moment, CEOs can choose to build VR tools that abandon politics in favor of gaming, design AR tools to educate on human rights, or make MR aimed at spreading biased news or disinformation. What will it be? The American people have essentially no say in this decision, but very possibly could be left with its consequences.

The Anti-Defamation League (ADL), which works to stop defamation of the Jewish people and other embattled groups, has suggested that social VR technologies will pose new challenges as they become incorporated into the social media landscape.[12] A report from the ADL outlines both the promise and the peril for VR as a new medium and also highlights recent cases of misuse, including incidents of virtual sexual harassment over VRChat and the game QuiVR, and the simulated physical harm during the Tribeca Festival's VR play *To Be with Hamlet*. The report highlights research from the firm The Extended Mind suggesting that large percentages of regular VR researchers, both female and male, have experienced virtual sexual harassment.[13]

People over retirement age and under the age of eighteen are particularly vulnerable to VR-driven propaganda, according to Joseph Sullivan, the founder of the California-based VR firm Luciton Virtual. Moreover, researchers from New York University and Princeton University have found that the elderly are especially likely to share false news. According to an article on their study: "Facebook users aged 65 and older shared more than twice as many fake news articles than the next-oldest age group of 45 to 65, and nearly seven times as many fake news articles as the youngest age group (18 to 29)."[14] My research at the Computational Propaganda Project at Oxford consistently revealed that young voters between the ages of 18 and 25 and older voters over 65 were particular targets of online disinformation campaigns during elections, security crises and natural disasters. The people we interviewed for the project, including political marketing consultants and digital mercenaries who had internal working knowledge of said campaigns, suggested that teenagers and retirees were targeted because they were seen as both especially vulnerable to false information and most likely to share it. Teens and young voters, interviewees told us, potentially lacked the critical thinking or media literacy skills developed in university or the workplace. The retired generation were not digital natives, they didn't know how to spot bogus news websites or suss out an automated profile. Despite this, they continued to sign up for and use sites like Facebook in droves.[15]

Sullivan's company is geared toward providing VR experiences for the elderly. He told me that he has been very clear with his clients that these technological experiences should not replace regular human interaction. Children's use of VR games and educational technology must also be explored. Though we know a great deal about how communication technologies like television affect both young people and older watchers, we know much less about nearly every other kind of more recent technology. The subjects of smartphone and video game addiction are relatively new by research standards; how long will it be before we have a meaningful understanding of the effect of XR media on the human brain? Almost by definition, it will be impossible to study useful data until long after the devices have been released and widely adopted.

Extended reality media also have implications in the realms of intellectual property, ownership, and identity. On many extant social VR platforms, users can dress up or be a version of whatever character or person they like—it's like one big online costume convention—but this brings up further questions: Who owns what in VR? How will those who create characters, whether in movies, books, or elsewhere, maintain ownership of their creations? The lines of copyright and intellectual property, which are less than clear in the first place, can be blurred almost infinitely by extended reality. Today there are message boards full of fan fiction and notices of cosplay conventions where people in homemade regalia create an experimental universe of different identities. Some people write completely new Harry Potter books, while others dress up as a young steampunk Gandalf.

It is one thing when people create such things at a convention or for the purposes of entertainment among a small community of ardent fans, but what happens when they spread to the virtual realm? One standard exception to traditional copyright constraints is the concept that any creation that transforms the original idea, song, character, and so on is not a copyright violation. Does a Gandalf-inspired wizard, created anew out of VR pixels, qualify? Do we need a new definition of "transformed" before we can answer that question? More perilously, what happens if the inspiration for the VR character comes not from fiction but from popular culture, or from politics? What happens if a virtual celebrity is depicted attending a racist march or making hateful remarks? Is it newsworthy? Should the VR space be regulated to ban or limit such activities?

Our copyright and free speech laws offer little guidance in such cases. And those who are building and regulating the extended reality space have not taken up the cause. But as this technology becomes more widespread we will run into a good many problems associated not only with perceptions of ownership or satire, but also with indoctrination and harassment. The present controversy over abuses on Twitter and Facebook and other social media sites offers a chilling example of what may happen if no rules are made. What happens, for instance, if thousands of VR users deploy hyper-convincing avatars of the US president doing or saying horrible things?

Perhaps nothing, if the strong free speech laws in the United States are enforced. But what if white supremacists in Germany create Hitler avatars and use them to spread anti-Semitic content or disinformation about an attack? What about people in authoritarian countries with strict laws monitoring defamation of leaders who use XR to mock those in power?

On the one hand, free speech is an important mechanism for checking power. Being able to openly criticize leadership, whether via XR or in the newspaper, is an integral part of any truly free country. The simple truth, however, is that free speech laws do not exist the world over. Beyond this, free speech can be used as a smoke screen by extremists pushing hate speech or propagandists selling false news. We have certainly seen this tactic used by US alt-right groups as they attempt to avoid being booted from social media platforms. With the spread of XR, though, uses may be even more challenging to track and parse.

Sensible XR companies will think about this problem before launching platforms where criticism, fake news, and harassment could occur. Strict terms of service and codes of conduct alone, however, will not be enough to curb misuses of VR or AR. It would be possible for firms to deploy AI technology—including amended versions of the generative adversarial network (GAN) technology that goes into the creation of deepfakes—to track avatars through reverse-engineering, using, say, Hitler's visage. But the perils of such ventures are the same as the perils of using AI alone to search for false news reports on social media. Humans, most likely assisted by smart automation, will have to play a moderating role in stemming the flow of problematic or false content on VR.

The Use of Virtual Reality for Social Good

Rev. Dr. Patricia Novick is the founding trainer of the Multicultural Leadership Academy, a multi-month training experience that brings together Latinx and African American leaders to learn about and collaborate on key issues facing their communities. Both Reverend Novick and I were lucky enough to serve as inaugural Belfer Fellows at the Anti-Defamation League's Center for Technology and Society. As fellows, each of us brought our experience and expertise to bear on a study of the ways in which groundbreaking technologies could be used to educate people about difference. We also looked at how digital tools could be used to challenge democracy. I set out to study how computational propaganda distributed over social media was being used to attack people of Jewish backgrounds and faith. I also worked to develop policy recommendations to pass laws to stop these hateful uses of social media. Novick, who also served as the first corporate social responsibility director at McDonald's Corporation and as a senior fellow at Harvard's Center for the Study of World Religions, focused on how extended reality could be used to foster community and connections.

Novick is currently working with the Chicago-based Bronzeville/Pilsen Augmented Reality Project, "a community-based project to develop a mobile phone 'app' using emerging technologies that highlights the positive cultural, historical, artistic, and natural resources within those minority communities (one largely African American and the other largely Latinx)."[16] The project uses AR to teach these groups about their homes and social groups. Importantly, it is aimed—like many of Novick's endeavors—at fostering connections between different and diverse peoples.

Other groups have set out to use extended reality media to educate people about groups they might misjudge, misunderstand, or simply never encounter. These uses of VR and AR can foster community and fight polarization rather than furthering social divides. The logic is that if someone can see, hear, and feel another person's lived experience, then perhaps they can become better able to empathize with their experiences. VR could, for instance, allow all types of people to experience the impossible literacy tests that African Americans had to pass after the Civil War in order to vote. It could let folks from all over the country share Rosa Parks's experience on that bus in Montgomery in 1955. It could let them walk alongside Dr. Martin Luther King Jr. as he fought for civil rights during the 1960s.

A team of researchers at the University of Barcelona reports that "spending a few minutes in virtual reality can change white people's perceptions of other skin colors."[17] According to the team, experimental tests using VR reduced people's inclination toward racism. Professor Mel Slater argues that the technology could be used in similar ways to combat racism over the long term. The activist technologist Clorama Dorvilias seeks to use VR to "gamify empathy" by revamping inclusion and diversity training.[18] Dorvilias argues that extended reality tools could be used to remove the shame from these kinds of trainings, which feature prominently in corporate onboarding and extended educational programs from Tokyo to Silicon Valley. She told *Motherboard* that "the nature of VR has limits to it. Until people try it, they won't really understand why it's so powerful. Once they try it, they're transformed."

If we can imagine VR- or AR-driven social media platforms, à la Kelly's "mirrorworld," then we should probably be asking ourselves how news will be incorporated into these future spaces. In fact, news organizations are already beginning to incorporate VR and other extended reality media into their journalistic practices. But decisions—or a lack of them—about how to convey news and reporting, or disinformation and propaganda, have been among the primary causes of the largest problems faced by today's social media firms.

The Reuters Institute for the Study of Journalism, a research entity based at the University of Oxford, argues that while VR is a promising new medium for reporting, "VR news still has a poor understanding of its audience both in terms of content, content discovery, and attitudes to the technology and hardware."[19] These issues, considered alongside the potential legal hurdles generated via the news or information that companies choose to prioritize for users, present both possibilities and perils for the future of these immersive technologies as news media devices. As the Reuters Institute points out, members of the news community must work together to make sure that users have optimum experiences when they engage with news via VR.

That said, it is crucial that these decisions be geared toward promoting fairness, accountability, and transparency as much as they are toward financial success for news organizations. For instance, news organizations might need to limit how they share their VR products to ensure that they are not copied, manipulated, or gamed. It would be terrible if a news outlet's VR platform could be emulated to, say, spread disinformation during a crisis. Moreover, VR cannot just be presented as a novelty by serious reporting outlets—it needs to contribute substance to how journalists share the news and how people consume it.

VR and AR can be used to transmit all sorts of other educational and scientific information beyond those already listed. They can even be used to combat problems like climate change. The island nation of Palau comprises 500-plus islands scattered across a remote piece of ocean in the Western Pacific. The small republic is home to just over 20,000 people of mixed Palauan, Filipino, Chinese, and Micronesian descent. Unsurprisingly, life there is intimately tied to the ocean. Fishing and maritime tourism, including scuba diving and snorkeling around the area's vibrant reefs, are among Palau's largest industries. However, life and commerce in the island nation are rapidly changing. Climate change is reducing precious land space. According to the *Independent*, environmental shifts prompted by human activity are also "set to affect Palau's waters, which could become more acidic, threatening corals and disturbing fish stocks already depleted by overfishing."[20]

Palau was the second country to sign the Paris climate agreement to abate the global rise of temperature. But without buy-in from world superpowers, including the United States, Palauans are having to get creative in their responses to the problems created by global warming and climate change. Education is a key component of helping politicians and citizens in the country make informed choices about how to protect their precious oceanic resources. Extended reality media is being leveraged in Palau to allow policymakers to experience what they face. According to an article on the topic in *National Geographic*, most locals favor fishing over scuba or other activities. It is tourists, often with little understanding or vested interest in the local reefs and wildlife, who spend the most time diving in and among the coral ecosystems.[21]

A research team from Stanford University, led by Dr. Jeremy Bailenson, wanted to show government officials what can happen when such activities go unchecked. "If a picture is worth a thousand words," said Bailenson, "then a virtual reality experience is worth 1,000 pictures." Using this logic, he and his colleagues created a VR experience that puts the user in the midst of a group of tourists unwittingly smashing a portion of coral reef with their diving fins. This destruction had actually happened, and the Stanford team filmed it and converted it to extended reality media. When members of the Palauan House of Delegates put on a VR headset and experienced the footage, they were aghast:

> *One suggested that the virtual reality simulation be mandatory for tourists before they board snorkeling boats so they know what not to do during the excursion. Another suggested that virtual reality demonstrations be put in schools. And legislators enthusiastically passed a Call to Action that day to integrate climate change research into its ocean policy, through "assessments based on scientific study and traditional knowledge systems to inform and support decision making."*[22]

In this case, VR was a powerful tool for inspiring positive political change. It allowed Palauans in positions of power to gain new knowledge about their precious and precarious natural environment.

VR and AR can also be used by artists to create beauty, foster compassion, and critique power. *The New Yorker*'s David Remnick set out to find the Orson Welles of VR but quickly realized that there isn't one just yet.[23] He points out that while VR could create a future of wearable movies, artists and others have not yet truly realized the storytelling potential of extended reality tools. Georgia Tech professor Janet Murray elaborates on this point: "Right now [VR is] just a technology. That is the equivalent of inventing the movie camera, but inventing the movie camera is not the same as inventing the movies."[24] Though there has been a lot of hype around these technologies, several logistical, technical, and creative barriers need to be overcome before their creative potential will begin to be realized.

The same can be said for the use of VR, or any other type of XR, for the purposes of propaganda. We do not yet know how computational propaganda will manifest itself on these types of platforms. We can look to known cases of deepfake technology and social media propaganda, however, to hypothesize about how disinformation and political harassment will flow using a medium like VR. It is likely that propagandists will attempt to exploit any public entry points for building software on social VR platforms. Users of any open API that enables people to upload their own creations must be thoroughly vetted. Even absent a feature that allows users to build software on top of these platforms—to create VR bots, for instance—there will still be efforts to spread poor-quality and potentially dangerous information using XR. Creators, regulators, and others must vigilantly monitor for this eventuality.

Slow Extended Reality

While filmmakers like Alejandro Iñárritu are beginning to explore extended reality media, they still haven't cracked the code for making an XR experience that truly draws the user in. But as time progresses, and as VR and AR tools become more affordable and more mobile, people will design VR that is as captivating as a normal film and as addictive as Facebook or Instagram, if not more.

Imagine a social world like the Oasis, portrayed in the book and film *Ready Player One*, where people live significant portions of their life in a simulation. They can meet, interact, even fall in love with people in this digital landscape. They can generate the life they have always wanted, be who they want to be, look how they want to look. But they also face new challenges not present in today's social media. They are immersed in an environment that feels, looks, and sounds real, and such an environment makes catfishing, the practice of hiding one's identity behind a virtual avatar, easier than ever. It may also enable any number of previously unimagined scams, such as entreaties for money to help escape a supposedly dangerous situation, the theft and ransoming of someone's VR properties, and who knows what other exploitations and potential abuses. Before these possibilities come to fruition, while we are still in the early days of VR and AR exploration, we have an opportunity to generate a sensible toolkit for dealing with problems like virtual propaganda, harassment, racism, and hate. We cannot continue our habit of addressing the problems that come up with our new technologies only years after they have been identified.

Rapid innovation in VR and AR have resulted in exciting social technologies that can help users understand circumstances outside their own experience—what it feels like, for instance, to experience racism, sexism, discrimination, and hate. But if VR can be used to fight racism, it can also be used to advance it. There's no reason that this tool can't be used to advance the cause of hate groups. Within the technology itself are some ways of combating these types of malicious uses. VR can be used for real-time social interaction, such as in a world like the Oasis. It can also be used, as discussed earlier, to put people in the midst of historical experiences, as well as to show them potential futures. The team at Stanford used VR technology to show Palauans how their natural environment was being destroyed in the present day. What if we used VR to show people what the world will look like in twenty years if we do not address global warming? What if we build AR tools that we could point at our natural environment to show current pollution levels or the endangered status of particular plants and animals? Could VR be used as a wake-up call for the ills that surround us? We could even create VR modules to train people on the perils of computational propaganda.

One thing is clear: we must treat the potential harms of extended reality tools as seriously as we treat the potential benefits. The ADL report highlights the fact that harassment is *already* taking place online. We must not treat such offenses lightly. There should be clear user codes of conduct for all social XR media experiences. If users fail to meet expectations, then they should be banned from platforms. If they threaten offline violence or consistently bully others, then they should face serious repercussions. Hate speech must never be tolerated. Disinformation and manipulative political campaigns, including those by automated bot-driven or cyborg XR users, should be banned.

Companies and platforms can mitigate misuses of extended reality media by abiding by several principles.[25] First, companies should have clear rules about transparency—especially on public platforms. They must require users to be transparent about their identity and be equally open about how this information is stored and shared. Twitter and Reddit have shown that anonymity on public social media, especially those where massive amounts of news flow, can lead to significant trolling and misinformation problems. When people know that anonymity will help them avoid facing repercussions for their actions online, they are much more likely to engage in malicious acts. I understand that anonymity has clear purposes, including for the protection of activists attempting to use digital tools to communicate in the face of oppressive regimes, but transparency around identity and anonymity are not necessarily mutually exclusive. A company could, for instance, privately vet users' identities while still allowing them to access and use their platform.

The company behind an XR platform intended to be used for protective democratic purposes—if users are not identified publicly—must still gather information on users to verify their identity. Social VR companies can even work with third-party security companies or trusted democratic governments to encrypt and store said data off their own servers. No matter how data is collected and stored, collecting identity information enables companies to hold users who violate the principles of the platform accountable. Verification before allowing anonymous use is even more necessary on platforms where people are not identified to one another by their real names, because it is on these platforms that governments and other powerful political actors often coalesce to try to target political opposition. By no means will transparency efforts around verification be simple, legally or otherwise, but they will ultimately serve to protect users and their experiences from malicious acts.

The principle of transparency on XR systems also compels companies to be clear about the user data that they collect, store, share, and sell. They must institute another principle, trust, in order to assure users that they are not misusing their data or unduly profiting from it. This is especially important if companies are gathering and storing user identity information, as in the case just discussed. Terms of service about the data that XR companies gather must be concise and easy to read. Society can no longer allow technology companies to circumvent transparency and user trust with impossible-to-understand terms of service agreements. Providing clear terms actually benefits the companies themselves, because such transparency makes users more likely to trust—and thus return to—their product.

Twitter can attest to the problems that arise when users have little trust in a platform and its policies. User growth stagnates, money is lost, and public opinion turns against the company. Even Facebook and YouTube must by now know this to be true. They can continue to race to spread their products to previously unconnected parts of the world in the same way that Philip Morris (now known as Altria) worked to spread tobacco products to Asia and other locales after smoking rates dropped in the United States and other regions.[26] But eventually the tale of lost users will catch up to technology companies in the same way it caught up to Philip Morris. No one will trust them and they will begin to fail. XR firms and platforms must take heed and build platforms that incorporate both trust and transparency as key, measurable principles, not just as token parts of a mission statement.

Another core principle for extended reality media firms, especially those with social functions, is coordination. Google, Facebook, Twitter, and other social media firms have seriously failed at working together to address the problems posed by computational propaganda and other platform misuses. The companies operate on an old-school model of total corporate opacity in an age when users constantly move from one digital communication platform to another. Disinformation does not simply stay on one platform—it spreads from one to another, staying in constant motion. While it is understandable that platforms want to protect proprietary information, it is also unthinkable that they haven't done more to coordinate with each other to prevent misuses. It is unlikely that there will be any winners-take-all, like the VR platform the Oasis or Kelly's AR platform mirrorworld. It is more likely that social XR will come in many shapes and sizes and be run and owned by many companies. The antitrust laws associated with maintaining democratic media systems should make this inevitable. With this in mind, tomorrow's social media companies should collaborate to prevent manipulation and hate from the outset. We cannot allow these technologies to break reality, regardless of whether or not they provide a "virtual" experience of it.

This brings up another point: virtual reality is just that, a simulation of life. There are two principles to follow that will prevent this technology from becoming irreparably addictive and unduly psychologically influential. First, companies should prioritize moderation, and I do not mean moderation as in content monitoring and general maintenance of platforms—though this is undoubtedly important. I mean that they should prioritize moderate uses of their platforms, which should be designed for healthy use. New social XR platforms should provide clear indications of when they are being overused. Numerous smartphone apps that track and limit overuse of technology, such as Social Fever and AppDetox, are already on the market. As with these apps, something similar could be built into new XR platforms, or over them, to make sure that users don't spend all their time within the digital experience. It may be challenging to enforce these rules, short of shutting down the platform for those who spend more than a certain amount of time on it, and users themselves will have to buy into them.

Technology companies should not prey, however, on users' attention. In fact, the rise of the attention economy (a framework through which social media companies monetize, and therefore attempt to maximize, the time we spend using their products) must be stopped if we are to preserve public health. Second, XR platforms must provide accountability for the experiences they offer. Fiction and fact must be distinguished, especially when users are attempting to consume legitimate news. Of course people will be aware that they are playing a VR game or engaging in some other VR fantasy, but as social XR progresses it will be more challenging to identify what information relates to the offline world and what does not. Companies must be accountable not only by passively flagging problematic data and fact-checking disinformation after it has gone viral, but also by building systems that track and stop inorganic political manipulation campaigns and coordinated trolling attacks before they gain steam.

Finally, XR platforms must operate via the principle of inclusivity. They should not only provide equal access to tools for all sorts of people, but work to give minority social, ethnic, and religious groups equity in their tools. To preserve democratic and heterogenous use of VR and AR it is crucial that companies design their tools with difference and diversity in mind. Politically marginalized communities—those who lack a voice in mainstream political discussions—are crucial to the success of any democratic system, both online and offline. We cannot rely on platform leadership or governments alone to see or speak for all peoples when it comes to designing the next wave of technologies. Instead, a wide range of stakeholders must have equally valued seats at the table when new media tools that prioritize equality and prevent manipulative uses are being built.

Humanlike Versus Human

Virtual reality and augmented reality are amazing tools for simulating the lived experience. Not only can we become our favorite movie character, but we can travel to places we've never been to and meet up with friends in a place of total fantasy. Both XR and AI technology are pushing the boundaries of what it means to be human as technology and people become ever more intertwined. Bots are beginning to seem more human and humans are beginning to seem more botlike. As time goes by, there will be even more mixing between what is human and what is machine.

As our tools become more like us and vice versa we will have to redraw the lines between what is real and what is fake. More importantly, we will have to rethink our perception of the truth. If a friend in a VR simulation is telling us the latest news, does it matter that they are a humanlike bot? What if a human-sounding bot calls us on the phone and tries to talk to us about politics? The more we map human characteristics onto technology the more challenging it will become to distinguish between reliable information and junk content. Research shows that we humans are more persuaded by other humans, especially when we hear them or see them. The age of real-looking, -sounding, and -seeming AI tools is approaching, however, and it will challenge the foundations of trust and the truth.

Chapter Seven
Building Technology in the Human Image

Thus the first ultraintelligent machine is the last invention that man need ever make, provided that the machine is docile enough to tell us how to keep it under control.

IRVING JOHN GOOD[1]

@Futurepolitica1

In the first chapter, I told you about a shady conversation during my first SXSW conference. What I didn't mention there was that the event was held during the height of the 2016 presidential primaries. Members of Hillary Clinton's and Bernie Sanders's campaign leadership teams were there. Barack Obama even swung by for an appearance and, by all accounts, stole the show. At a DNC fund-raiser held during the conference, the former president, according to the *Daily Beast*, had some choice words about Trump. He said that the then-candidate was "a distillation of what has been going on in [the Republican] party for more than a decade."[2] Fittingly, I was there with some friends and colleagues to talk on a panel entitled "RoboPresident: Politics in an Algorithmic World."[3]

The discussions during the panel were a harbinger of things to come. We talked about the political bots we had already noticed being used in attempts to control communication online during the race. We discussed Microsoft's failed experiment with the Twitter bot "Tay" and talked about the use of bots to both support and undermine journalism. We even got questions from social media company employees in the crowd. We said that it was only a matter of time before AI technology was more effectively encoded into social bots. Our digital automatons would become more like us, we hypothesized.

When I was approached by a curator from the Victoria and Albert Museum about the project that would eventually become @futurepolitica1, he had a specific question: As bots become more human, will they be more effective at targeting politicians with propaganda? At the time, I was writing a piece on the US election for the book *Computational Propaganda* in which I reported on the discovery by my colleague Doug Guilbeault and me that even rudimentary political bots were fairly and significantly successful in getting US politicians to repost or comment on their content on Twitter. Some of the stuff the politicians and pundits reposted was even mis- or disinformative. The curator asked me if I could construct a political bot that showcased, for educational purposes, how these humanlike digital automatons worked for an upcoming exhibition on future technology.

We got back in contact a few weeks later, when I arrived back in the United Kingdom for work. Because I am not a coder by trade, I invited Alex Hogan, founder of the Etic Lab, to join me in creating the bot. We purposely built the bot to be more informative than invasive. It shared information about political bots on its feed and could even interact with other Twitter users. People at "The Future Starts Here" exhibit could send it messages and watch as it responded onscreen. The bot would also ask different Twitter users for their opinions on tech and political issues. But it did not use sophisticated machine learning in these interactions, despite Alex's assurance that he and his collaborators at Etic could design something more "intelligent." The potential for the bot to go rogue or get tricked into saying something horrible, as happened with Microsoft's Tay, was too great. We created the bot to be humanlike, but not too human. It would openly tell the people it talked to that it was a bot built to do research and teach.

What was clear, even in 2016, was that political bots could be built using fairly sophisticated AI. It would have been much more expensive and time-consuming, but it could have been done. Today this is even more true. And as I mentioned earlier, it would also now be easier to scale from one ML bot to a group of them. These tools could be used to trick an unsuspecting politician into sharing disinformation, even potentially harmful information, about nuclear arms, health care, or foreign affairs. In Hawaii, for instance, serious hysteria quickly broke out when a mass-text message stating that a nuke was approaching the islands was accidentally sent to all residents; though the text was quickly reported to be an error, there was no bomb. What if bots were able to produce a similar effect over social media?

We are making our machines look, sound, and act more like us, and the ramifications extend well beyond bots. AI is the story not merely of a new technology but of its convergence with human behavior.

Machine or Mortal?

Although many experts maintain a healthy skepticism about the nearness of the singularity, the age of artificial general intelligence (AGI), it is undeniable that the line that separates people from machines becomes blurrier with each passing day. We may not have empathetic machines that exercise free will, but we are creating increasingly sophisticated technology that can learn from its environment. Artificial intelligence, of course, is at the foundation of these advances. Machine learning, deep learning, and their various hybrid combinations and subdisciplines are also deeply focused on ways to make our tools smarter and more efficient. As *The New Yorker*'s Tad Friend points out, we are becoming oblivious to AI-powered tools: "A Yahoo-sponsored language-processing system detects sarcasm, the poker program Libratus beats experts at Texas hold 'em, and algorithms write music, make paintings, crack jokes, and create new scenarios for 'The Flintstones.'"[4]

Beyond this, though, AI technologies are actually allowing us to build tools that look, speak, and even feel more like us. It is one thing to have a card-playing robot that can act like a person, but it is fully another to have one that looks and speaks like a human. Voice recognition and speech generation technology is beginning to sound more and more human and less and less robotic. Digitally generated images of the human face are becoming difficult to distinguish from the faces of real people. We have social bots that can learn from conversations and hold lengthy conversations with human users. Many of the physical robots we build look, move, and speak like people. They are even programmed with humanlike tics and speech modulation.[5]

Over the course of my work in this space, I have talked to and formally interviewed many empathically minded engineers and inventors. Many of them state that their explicit and primary goal is to create technology that makes human life easier: faster internet, cars designed to prevent accidents, higher-definition photographs and film, better ways of compressing data, or drones designed to monitor climate change. These tools computationally enhance the speed of jobs that would take humans, on their own, much more time. They do things that people, without digital tools, could not achieve. Sometimes these tools are even built to do things that we simply do not want to do.

The implicit desire of many of these technologists, however, is less humanitarian. They are often driven by simply a need to innovate, progress, and create. These are not bad motives, but as we continue to design tools in the human image and with humanlike characteristics, we must be increasingly cognizant of the ethical implications of particular uses of these anthropomorphic tools. We must also be aware that the ways in which we build technology have implications for how it transmits facts, truth, and reality. The more human a piece of software or hardware is, the more potential it has to mimic, persuade, and influence.

Building Relationships with Machines

As a longtime Apple user, I have gotten more and more used to Siri. Like almost everyone else, I have a tendency to personify the digital personal assistant. Some people yell at Siri or play at trying to have deep conversations with her. Apple understands this and has programmed Siri to respond cheekily to particularly rude queries and with commensurate bluntness to unclear ones. Siri is programmed, in factory settings, to sound like a woman. I have already unthinkingly referred to "her" in the previous sentences. A number of researchers and publications have aptly pointed out that applications like Siri, Microsoft's Cortana, or Amazon's Alexa "sound like ladies because of sexism."[6] These software-driven assistants reify historical sexism. They reproduce tropes of the woman as a secretary, a maid, or a nanny. The preprogrammed voices of these products sound like women and, as time goes by and technology progresses, they sound more and more human.

The way we build our tools and the qualities we map onto them matter for a variety of reasons. Some experts and engineers think that if people perceive a tool to be humanlike, or to convey human qualities like politeness, they will treat it more humanely.[7] On the other hand, if we think a tool is more robotic, they argue, we treat it in a more utilitarian fashion.[8] Most people have had the experience of feeling angry with an automated telephone system, especially when it refuses to pass us on to a real person. Some of us even yell at the system, while others have figured out that, by swearing at certain automated operators, they quickly get passed on to a human one.[9]

It is possible that our interactions with technology will change as that tech becomes more like us and less like machines. The manners of the AI voice on our phones may become more polite or less polite, more varied or funnier. It's hard to know for sure, and depending on the outcome, we may become more or less able to manipulate machines—and they to manipulate us. As this ecosystem continues to evolve there are several questions we must reckon with to make sure that, in our quest to create ever more useful technology, we do not at the same time build tools that are ever more effective vessels for manipulation or subjugation.

How might the next generation of humanlike AI make people more confused, more sexist, and more angry? How might humanlike digital disinformation campaigns be used to prey on the poor or marginalized? Also, will more humanlike propaganda be more persuasive? If it is—and the research suggests that this is likely to be the case—how will we stop it from falling into the wrong hands? The outcomes are far from guaranteed, but there certainly seems to be potential for misuse. With recent computational propaganda campaigns on social media, technologists and the general public did not know the answers to similar questions until well after the critical decisions about design and deployment had been made. We need not accept this as the new normal with our future tools.

As Grady Booch suggests, we must bake the best parts of humanity into our humanlike tools if we are to avoid malicious uses of AI or circumvent some version of an inhumane super-intelligent technology far down the road.[10] This being said, if we make our tools more like us, then could they not be more adept at tricking us with false information? Could they spread even more potent harassment or target those vulnerable to persuasion even more effectively by making it appeal to us through empathy or sympathy? And couldn't people deploy these personified digital servants in much the same way they use social bots—to sway votes and quash dissent on their behalf? While it is important to build our tools with our own highest virtues in mind, it is not clear that these virtues need to be emulated in a humanlike way by our machines—only that they should be beneficial to humans, human society, and the natural world in the long term.

We must also be careful that the good intentions of engineers do not lead to bad outcomes. For instance, it might seem like building humanlike voice technology for our smartphones would generate sympathy from human users and therefore allow more effective communication. But how could this technology be misused in the long term for the purposes of propaganda or other forms of manipulation? Could Google Assistant become another tool for spreading dangerous disinformation if someone figured out how to game its algorithm? Google set out to "do no evil." But such maxims do not work if designers don't think about future problems before launching their technology. Maxims can also be hard to abide by as companies grow.

Beyond Humanlike Audio

A slew of physical, offline robots have been built, and many of them have acted, moved, and talked like humans. But imagine that there is finally a robot that looks identical to us—let's call it MeBot. It has realistic features, speaks exactly like us, and has dance moves that would impress Beyoncé. It is not super-intelligent. The uncanny valley effect, the cognitive response which provokes the unsettling feeling that something is close-to but not-quite human, is still marginally there. It can, however, pass for a person if it doesn't have to have a long conversation. Best of all, you can send the builder's company photos, video, and audio clips and they will make the robot look and sound just like you.

Advances in AI and cheaper hardware have made it possible to buy a MeBot for under $5,000. The designers claim that, using the information provided by the buyer in an extensive online form, they can give the machine some degree of a given person's characteristics. People regularly use their MeBots for all sorts of things. They send them to run simple errands, like walking the dog. Some people even allow the automatons, which can be encoded with something called Nanny-ware™, to escort their kids home from school. Most of all, though, people keep them around for enjoyment, sometimes using the robots to play pranks on their friends. There is something super odd and funny about having a robot doppelgänger around. People almost seem to develop loving relationships with their MeBots—they just look so human.

One day, though, an anonymous group hacks the code behind a large group of MeBots in Ohio. No one notices, because the hackers allow the bots to continue doing their regular tasks. But it is voting day—a presidential election. Based on the constant stream of ambient data collected by the MeBots, the hackers can tell if each corresponding human is likely to vote that day. The hackers can also determine optimum times for the MeBots to "go rogue" during the course of their daily errands. That day, in one pivotally important Ohio county, 12,000 MeBots show up to vote in place of their humans and the high voter turnout prevents anyone from noticing the spate of quiet would-be voters.

The election that year is close. Later people will compare it to the 2000 Gore–Bush debacle, which was decided by just a few hundred votes in Florida. This time the outcome depends on the perennial swing state of Ohio. Later, after authorities discover that the few thousand MeBots voted, it is determined that they might have decided the outcome of the election. How did the election monitors, everyone asks, not notice the robot voters? Why, they ask, weren't the MeBots built with some kind of tattoo or corporate mark to show that they weren't people? After all is said and done, the Supreme Court upholds the election results. Publicly, the authorities state that they are confident that, according to the data, the outcome would have been the same without the robot votes. Privately, they have no idea how to litigate a bunch of MeBots voting. Some of the robots' owners, embarrassed, even claim to have showed up to vote—despite evidence to the contrary. It's a total mess, and the people who did the hacking are never found out.

This hypothetical situation may seem a little out there, but it serves the purpose of showing how humanlike tools could be misused to game politics and social life. Although the human voice can be a very powerful tool for persuasion, this imagined scenario makes it clear that it is not just voice-driven technology, like Siri or Cortana, that we should scrutinize.

We experience the world with five senses, and technology can play to all of them, not just hearing. In what ways might the incorporation of sight, taste, touch, or smell into our tools challenge our perceptions of what is true or what is worthwhile? How will reality be altered by robots that look like humans? What if they are covered in skin or hair that feels real? What will be the psychological effects of building tools preprogrammed to transmit scents at certain times of the day that remind us of when we were young, or our first date, or our first day of college? What about a tool that imbues foods with similarly evocative tastes? How could such technology be used to make our lives better? How could it be used to coerce or control us?

There are many technologies that are just arriving on the scene, some that are on the cusp of being created, and others that still exist only in our imaginations. How a tool is used comes down to human decisions: how a person designs it, builds it, uses it, or directs it. Social bots, for instance, can be used to critique power by calling out politicians for corruption over social media, but they can also be used to cement power by disenfranchising opposition voters with digital disinformation during elections. The tools of the future will not be exceptions to these rules. If we want to build technology that we are proud of—that actually can be used to support human rights, equality, and freedom—then we must think before we build. I am optimistic about the future of technology because we can make decisions now that will make tomorrow better. We cannot do this, however, by mistaking tools for people, or vice versa.

The Human Voice and the Ability to Persuade

During my research, I spoke to Dr. Juliana Schroeder to find out how language affects the ways in which people express their own mental capacity and evaluate that of others. Schroeder is an assistant professor at the Haas School of Business at UC-Berkeley and cofounder of the Psychology of Technology Institute. She does research into how different uses of language make someone seem smarter, more capable, or more humanlike.[11] She also explores, conversely, how other language uses can lead to dehumanization in another person's perspective. I asked her several questions about these effects: Can how we speak affect how intelligent people think we are? Can technology that sounds human be experienced more empathetically? What might this mean for the next wave of technological propaganda?

She told me that her work has led her to study the various mediated uses of language that occur through technological systems such as social media: written posts or articles as well as speech or audio, and combinations of the two, like audio clips with subtitles. Schroeder looks primarily at whether language is communicated verbally (spoken) or nonverbally (written). From there, she studies which type of usage is perceived by study respondents as more humanlike. Her findings provide clear indications of what types of language use result in humanization versus dehumanization.

According to Schroeder, people tend to "systematically believe a communicator is more intelligent, competent, and thoughtful" when that person communicates out loud—when they use their voice. Conversely, when respondents silently read a communicator's written post, they are more likely to see that person as incompetent and unintelligent. Interestingly, when Schroeder and her colleagues had respondents read such written posts out loud, they would read in a mocking or silly voice, especially when the post contained views with which they disagreed. But when respondents heard someone else read a previously written statement out loud—one that had clear markers of having been written first—they were much more likely to see the other person as thoughtful and rational. So what is the important factor here?

According to Schroeder, it is not just about hearing something versus reading it—it is about hearing a human voice. The human voice, she said, "moderates the effects of dehumanization." But, she cautioned, a respondent needs to be hearing a natural human voice in order for that voice to convey humanlike qualities. The voice must contain variance and natural rhythm to be seen as intelligent. A stilted or unnatural robot voice, in contrast, will not be heard sympathetically by respondents. But what happens when a robot voice is made to sound human? Over time, for instance, Apple's Siri has become more and more human-sounding—for example, by putting word stresses in the right places.[12] New tools, including Google Assistant, are now programmed to be similarly and, for some, shockingly human.[13] In such cases, Schroeder said, respondents still think that the message they are hearing is more intelligent and more sympathetically humanlike than a written post.

Schroeder's findings, of course, have important implications for how automated voice technologies might be used to manipulate public opinion. Since humanlike robot voices are mapped by listeners as rational and smart, these same automated voices could be used to scale attempts to engineer political conversations and public perception of candidates or causes. And machine-learning systems designed to convincingly respond to our queries could take such a development a step further.

Consider the political robocalls of days past. The minute most people heard the robot voice, they hung up, regardless of the message. As time and technology have progressed, however, political marketers have grown wise to this behavior. Now, they deploy prerecorded human voices that make convincing linguistic errors. The voices on these newer marketing calls pretend not to hear you say hello at first and respond with "Hello... are you there?" I myself have fallen for these calls more than once. But even though these human-voiced robocalls are more natural, they are not built to interact with the caller. When the "person" on the other end of the line steadily goes on and on regardless of, and oblivious to, your questions, you catch on—and hang up—pretty quickly.

Machine-learning technology, coupled with recent innovations that allow automated voices to sound more human, could change this response. For instance, more complex online chatbots can hold a realistic conversation with a human user for several lines. Such bots have even passed the Turing Test, which analyzes whether machine-borne communication exhibits intelligent, human-seeming behavior.[14] This test is often deployed online to keep bots off particular websites. But the conversations or prompts that test for humanness on the internet are generally written or visual, occurring over Twitter, a chat app like Telegram, or via online customer service chat for banks or other services. The bot, in these cases, has limited capacity for responding, even if it is built to learn from conversations and adjust its methods of communicating.

In my experience, even the most sophisticated chatbots have robot tells after five or six sentences. They therefore struggle to convince people of much on their own. Given Schroeder's research, they are also limited in not being as likely to be seen as rational or intelligent—and thus convincing—because their communication is written. This is exactly why many political bot makers build their Twitter botnets to bluntly barrage activists with the same spam or threats and work to trick trending algorithms into believing a particular hashtag is popular through sheer numbers. Forget nuance—simplicity alone achieves their ends.

As machine learning progresses, and as robot voices and bot-written text continue to sound or seem more humanlike, marketers will catch on to the effectiveness of these tools at swaying opinion. This innovation may still be several years out, but even in its current form it could be used to trick the young, the elderly, or those who live in countries with limited access to diverse media and technology. Could automated voice systems be used right now to manipulate such people over the phone or when they use other audio-based digital communication tools like Skype or Zoom?

Google's Duplex tool, a digital assistant that can place restaurant reservations and serve other humanlike functions, sounds almost too humanlike. Hearing it for the first time is more than slightly creepy, given that most people who interact with it know it is an automated voice. It can be used by people to make lifelike calls to others. Many in the journalistic and academic communities pointed out potential misuses of this sort of technology soon after Duplex's prelaunch event in the spring of 2018.

According to the *Guardian* reporter sent to the event, "The robotic assistant uses a very natural speech pattern that includes hesitations and affirmations such as 'er' and 'mmm-hmm' so that it is extremely difficult to distinguish from an actual human phone call."[15] The tool's uncannily human voice isn't the only problem, though. According to the same article, during a demonstration in which Google showcased Duplex making calls to other people, "the virtual assistant did not identify itself and instead appeared to deceive the human at the end of the line."

Though the company says that this feature might change, the absence of signals that identify the speech of the tool as robotically generated is indicative of a larger problem with "smart" systems designed to be like people. They are often not transparently presented as being automated. This is especially concerning when they are used in a social capacity. One of the main problems with political bots, for instance, is that they mimic real people without ever notifying other users that they are not actually human. In this capacity, they are used to trick trending algorithms into thinking that they are human and that their content should be counted toward trends or other uses. Like the bots built to pass the Turing Test, they are deceiving the very systems designed—at least in part—to keep them out. We face the same problems with similar automated tools, like Duplex, that are made to sound just like people.

In one of the example calls that Google showcased Duplex making, the program called a hair salon to make an appointment. The person on the other line showed no signs of knowing that they were not scheduling the haircut with a real person. This particular deceit may not be a particularly serious offense in the grand scheme of things, but it is easy to imagine circumstances where similar tools could be used for less benign means. What if these tools are leveraged for extortion, including scam calls, or for politically manipulative robocalls?

When I first started studying computational propaganda, my colleagues and I often referred to it as social media push-polling, or we discussed it as an online version of robocalling. We were using the media we knew and understood to describe the new thing we were witnessing. In the first waves of computational propaganda that I studied, bots would be used to manipulate people into supporting a candidate, often by attacking the opposition. When using phone-based push-polling, political campaigns or their paid marketing consultants would call a voter and present a bunch of attacks on the opposition in the form of survey questions. They would ask, for instance, "John Smith is a known liar and cheat who has used shady business practices. Will you vote for John Smith?" Some of the phone calls were made to independent voters using automated systems.

It turns out that this early comparison was not far off the mark. In fact, in the coming years we may see automated voice systems being used to game elections. It could look like something of a hybrid of push-polling and computational propaganda. In such cases, though, the automaton being used to manipulate voters will sound and behave like a person. As Schroeder's research has revealed, human voices can generate empathy and increase our perceptions of a communicator's intelligence. What would it look like to use tools like Google Assistant to manipulate people into voting for one candidate over another?

Voice recognition technology is another tool that allows machines to learn from interacting with people. Machine learning and voice recognition go hand in hand. It is now possible to build artificially intelligent software systems that can listen to, learn from, and make use of spoken language. These tools, built to understand speech patterns, could be harnessed for any number of future social uses. Google's popular "Translate" application is one such technology that is incorporating voice recognition into its functions.

People who use Translate will be aware that the tool has consistently gotten better at translating from one language to another, though for some languages more than others. The tool is getting even better now that users can get translations using speech and audio. Google has created new headphones, called "pixel buds," that translate almost simultaneously.[16] According to the *MIT Technology Review*, which highlighted the new hardware tool as one of 2018's breakthrough technologies, the pixel buds work fairly simply:

> *One person wears the earbuds, while the other holds a phone. The earbud wearer speaks in his or her language—English is the default—and the app translates the talking and plays it aloud on the phone. The person holding the phone responds; this response is translated and played through the earbuds.*

If you have a phone with a microphone, you can converse with another person along with Google Translate and will no longer wish, says *Tech Review*, for the instant-translation powers of the yellow Babel fish from *The Hitchhiker's Guide to the Galaxy*.

Phones and software applications are now even being designed to understand accents. If you speak Spanish with an Ecuadorian accent, the phone will understand, and even learn from, your speech. We can use this technology to break down barriers, to avoid potential misunderstandings, or to more efficiently communicate across cultures. But technologies built to understand more nuance in language, from accents to patterns of speech, can be useful in ways beyond just translation. If a person is more likely to believe someone who sounds and speaks like them, why not program the human-sounding automated assistant to speak with the same accent or in the same vernacular? What if a propagandist could make calls using Google Assistant in any language, with any accent, down to regionally specific ways of speaking? How about if foreign government actors, like the Russian Internet Research Agency employees who manipulated the 2016 US election, could avoid detection by using translation services to not only write in convincing English but also to speak in it?

Over time and as technology progresses, programs built to write and sound like us will inevitably become more humanlike. Advances in the field of artificial intelligence will allow these things to mimic us in ways previously consigned to science fiction. The devices may not be truly intelligent, and they are not likely to act on their own volition or to feel as humans do, but they could be potent tools in the hands of those working to manipulate and deceive.

Such AI applications are not the only new technology that allows for creating digitized anthropomorphic representations. Another is software that can create a realistic likeness of the human face.

Digitizing the Human Image

Nvidia is a technology firm that specializes in designing graphics processing units and chip units for mobile phones and cars. A 2018 report from the company has a lot of technology experts on edge. Why? The chip maker has created an AI-powered program for producing incredibly real-looking, machine-generated human faces. The team at Nvidia used a combination of generative adversarial networks (GANs)—which you'll remember from the chapter on deepfakes—alongside "style transfer": using AI to mimic the style of particular images or paintings.[17]

The photos generated by the new technology from Nvidia have all sorts of implications and potential uses. As *Motherboard* notes:

> *While this no doubt raises the specter of rampant AI-generated images fooling us into thinking they're real, it's worth noting that pulling this off took a week of AI training on eight Nvidia Tesla graphics processors that cost thousands of dollars each—not something you find in your average gaming rig.*[18]

But as with most technology, these processors are likely to become cheaper as demand increases and the tools progress. So how could realistic-looking, machine-created human faces be used for the purposes of computational propaganda or other nefarious practices? For one, those building proxy social media accounts for manipulative purposes would no longer have to steal other people's profile pictures or use random images. After all, reverse image tracking and other forms of profile picture analysis are key methods for catching inorganic information operations.

When my team at Oxford began doing research into the use of computational propaganda and political bots during the Brexit referendum and the 2016 US election, we noticed an interesting pattern. Many of the profiles that we thought were bot-driven or otherwise engaged in spreading disinformation or political attacks used the same profile pictures over and over again. This pattern both tipped us off to their likely inauthenticity and provided a valid reason for reporting the malicious accounts to Twitter in order to get them deleted.

In other circumstances, we would suspect that the profile picture of a particular Twitter account or Facebook profile was a stock image or in some other way not representative of the other content on the page. Using Google's reverse image search, we would see if the photo was tied to other content online. Many times we would quickly find out that the account was run by a propaganda operation and using either the stolen image of a real user or a stock photo. AI-generated human faces would render this method of sleuthing useless, though there are probably other more sophisticated means of identifying such images using tech similar to Nvidia's.

Another problem with AI-generated human faces is that they could be used for online harassment while also working to manipulate how people, and which ones, share information related to a particular topic. In the past I've seen extremist political groups and racist organizations attempt to trick people into believing disinformative content about a situation or another social, ethnic, or religious group by impersonating members of the group they are attempting to manipulate (but to which they do not themselves belong). In another case, I saw individuals impersonate Jewish people online in order to both support and attack Israel, a tactic designed to sow division in the community.

The impersonation tactic is effective for a simple reason: if others believe that these invaders are part of the in-group that they are critiquing, or part of the in-group and critiquing another group, then they are more likely to view them as a reliable or unique source. For instance, someone perpetuating this type of manipulative astroturf political campaign could use digitally generated profile pictures of African Americans to invade Black Lives Matter conversations on social media with less risk of detection by researchers and tech firms. They could engage in conversations about women's reproductive health issues using a fake photo of a woman that is both untraceable and indistinguishable from an actual human face.

Finally, these AI-generated images could be used to damage the way we treat photo-based evidence in the judicial system.[19] In the same way that deepfake videos use the faces and bodies of real people in order to make them do or say things they did not say or do, so too could artificially generated faces be used to challenge perceptions of what is true or real. People could paste a fake, untraceable face onto the body of someone committing a crime. This image could be put forth as evidence that a potential suspect was not at the scene of the crime, engaged in criminal activity—this other (fake) person was. Beyond this possibility and the other problems with fake profile pictures just discussed, such automatically produced images will do more to challenge the public's general trust in the veracity of online content than they will to reinforce that trust. And AI-generated images will help people operate anonymously in newer, ever more sophisticated ways.

The reverse side of digital image generation—facial recognition technology—is also a concern for those who worry about privacy, intellectual property, and other ideals. In the wrong hands, this technology, which uses software to identify unique human faces in images or video, can be harmful to both the truth and our general well-being. In recent years there has been a great deal of conversation among experts, pundits, and the public about the perceived benefits of facial recognition. Governments argue that they can use these tools to identify terrorists or criminals attempting to cross into their country. Indeed, many airports now use facial scans as a means of verifying international travelers' identity as they pass through customs. Militaries and law enforcement suggest that they can use similar tools in the service of public safety, to identify an adversary from a remote location using a drone, or to catch criminal suspects by matching their faces to footage from a crime.

Yet there has also been significant dialogue among multiple sectors about how facial recognition technology could be used to violate people's privacy. For one thing, governments can use it to spy on their own people—which is more or less what the airport scanning policy does—regardless of whether they have previously been identified as criminal or military threats. And access to the technology may not be limited to governments. Could nefarious actors, from organized crime to known terrorist groups, use it to identify future victims? Beyond these concerns, how do we manage consent in the use of these types of tools? Where do we sign to allow automated cameras empowered with smart software to track our faces? These questions tend to evoke images of Big Brother and fears of the police state.

I understand both sides of the argument, though in my view the threats are greater than the promises. Meanwhile, as someone particularly concerned with disinformation and propaganda, I see other potentially harmful uses of facial recognition tools. One such danger is closely related to government spying: foreign adversaries or other powerful political actors could use facial recognition to identify targets for state-sponsored trolling campaigns or other vicious digital political attacks. In late 2018, Facebook encountered just such a plan. They told the *New York Times* that they had deleted over sixty accounts associated with organizations of individuals who were building and using facial recognition tools on behalf of the Russian government.[20] In a cease-and-desist letter to several of the firms in question, Facebook said, "Facebook has reason to believe your work for the government has included matching photos from individuals' personal social media accounts in order to identify them."

Once identified, targets of state-sponsored digital attacks would have a very hard time escaping their persecutors. In addition to their social media accounts and online personal systems, victims could be located at any number of important locations like banks, airports, and hospitals. The reach of this kind of surveillance could be near-total.

Upon digging deeper, the *Times* reported that the Russian firms had been operating on Facebook and elsewhere for over four years. This news, alongside acknowledgments from some of the firms that they also scraped Google image search for Kremlin facial recognition purposes, suggests that these entities were able to gather a great many images from citizens of all manner of countries. Given the strict eye that Russia keeps on journalists, dissidents, and academics, it would not be surprising to learn that it was using such images to build a database of "people to watch." Among the companies booted from Facebook for such activities was SocialDataHub. As the *Times* notes, the banner on that company's website reads, "we know everything about everybody."

In the summer of 2018, another socially and politically problematic facial recognition story emerged: Amazon's facial recognition software wrongly identified twenty-eight lawmakers.[21] According to the American Civil Liberties Union (ACLU), the company's "Rekognition" tool incorrectly matched US politicians' photos to mugshots of criminals. Of perhaps even more concern was the fact that the mismatched images were "disproportionately" of politicians of color.[22] Was bias baked into the facial recognition technology? The ACLU seemed to think so, and many experts agreed. Research from the MIT Media Lab and from Microsoft actually found that commercial facial recognition software was often less accurate for people of color and women.[23] This suggests that the tools were either trained using only white faces or simply poorly built and vetted before being launched. What happens, after all, when someone not in a position of power is incorrectly tied to criminal behavior using facial recognition software? The ACLU and the media might not take notice, and that person could end up in jail.

Making Machines Play Friendly

There are several things that regular people—as well as tech firms, politicians, activists, and journalists—can do to make sure that humanlike tools such as automated voice systems are not used to manipulate public opinion or further the causes of those looking to do others harm. Going forward, we can also study what we already know and how we are currently experiencing computational propaganda to protect those most vulnerable to disinformation and other means of challenging the truth. A great deal of the knowledge we already have about how to track digital disinformation campaigns online, especially the research on detecting social bot networks, is pertinent when it comes to preventing misuses of future anthropomorphic AI technology.

As with any automated technology with social uses, it is crucial that the technological creations of designers and engineers explicitly state that they are automated or generated using artificial intelligence or machine-learning technology. Twitter now requires that anyone hoping to build and launch third-party software through its platform's API be registered. This has prevented a great deal of malicious and spam-oriented bot use on the Twittersphere. Human-sounding voice systems should be similarly compelled to openly notify those they call or communicate with that they are digitally derived. Political campaigns and companies hoping to use these tools for campaigning or commerce should be made to register their intent with the FEC, the FCC, or the corresponding regulatory entity outside of the United States.

Companies like Nvidia should be required to devise ways of tracking or detecting whether an image is machine-generated. The use of computer-made faces should be strictly regulated so that propagandists and those hoping to fabricate evidence do not harness the software for such means. Facial recognition, though already on the market and in large-scale use by governments and companies, must be reconsidered in light of not only privacy concerns but also its potential to help malicious actors track or harm individuals with whom they do not agree or whom they do not like for other reasons. Biases in this technology and its potential misuses in the criminal justice system have been proven, and they must be mitigated. People must also be given the right to consent to have their faces scanned.

Beyond what politicians and tech company executives should be doing, regular people can simply be careful about what photos and voice clips they share online. Everyone knows that if you wouldn't want your mom or your boss to see it, then you should not put it online. Fewer people, though, are aware of the need to be careful about giving up information or images of themselves because doing so could make them a victim of politically motivated trolling, disinformation, or any variety of scams. A version of your face or your voice could even end up being used in a deepfake or as the inspiration in a political bot profile. When regular people protect their own data, they protect both themselves and perceptions about what is true about them.

Chapter Eight
Conclusion: Designing with Human Rights in Mind

Finding solutions to the problems posed by online disinformation and political manipulation is a daunting task. The digital information landscape is vast, and it extends beyond our current ability to track or contain it effectively. Moreover, the internet grows larger every day. According to a 2017 report on the state of the net from software firm DOMO, which combined in-depth research from multiple companies and news outlets, we create 2.5 quintillion bytes of data every day. Moreover, the number of internet users grew by one billion (to a total of 3.7 billion active users) in the five years previous to the report.[1] A 2018 *Forbes* article asserted that 90 percent of the online data available in the world was generated in the previous two years.[2] This means that the people working to game public opinion or exert social and political oppression using online tools have almost unimaginable amounts of data available on potential targets, with new information beaming out to them every millisecond. They also have access to a lot of potential targets and can leverage online anonymity, automation, and the sheer scale of the net to remain nearly untrackable.

Important ethical and legal considerations, along with the near-impossibility of finding a skillful operative, make prosecution a poor strategy for stamping out computational propaganda. Instead, we must fix the ecosystem. It's time to build, design, and redesign the next wave of technology with human rights at the forefront of our minds.

Thinking about responses to the rising tide of computational propaganda, I find it helpful to break them down into responses for the short term, the medium term, and the long term. Because of the quixotic nature of technology today, I usually consider tool- or technology-based responses the shortest-term fixes of all. Many of these efforts are Band-Aid approaches focused on triaging help for the most egregious issues and oversights associated with the infrastructure of Web 2.0—the internet of social media. Such amendments include tweaks to social media news algorithms, the code that identifies trends, or software patches for other existing tools. They also include ephemeral new applications for identifying junk news or browser plug-ins that track and catalog political advertisements. These efforts are useful as far as they go, but online manipulation tactics are constantly evolving. What works to track disinformation or bots on Twitter today might not be very useful a year from now. In fact, many of the applications that have been built for such purposes quickly become defunct owing to code-level changes made by social media firms, a lack of funding or upkeep, or propaganda agents finding a simple way around them.

There are useful products of this kind, like BotCheck and SurfSafe from RoBhat Labs; these detect computational propaganda on Twitter and check for fake news via one's browser. But these programs need to be constantly updated and translated to other platforms to stay relevant and useful. They present a promising start for tools that alert users to the threats of disinformation, but they must be combined with action from technology firms, governments, news organizations, and others in order to truly be effective. Another example of a propaganda tracker is the Hamilton 68 dashboard, a project from the Alliance for Securing Democracy at the German Marshall Fund (GMF) that was built to track alleged Russian Twitter accounts.[3] Although it is important to identify and report nefarious or automated social media traffic, and equally important to notify users that they may be encountering false news reports, these efforts are too passive and too focused on user-based fixes to counter computational propaganda in the future. Also, it is important to remember that a good deal of research shows that post-hoc fact-checks do not work and that social media firms are engaged in a constant battle to catch and delete new and innovative types of bot-, cyborg-, and human-based information operations.

More than anything, I want to communicate that all is not lost. Not only researchers, policymakers, and civil society groups around the world but also technology firms are fighting to stem the tide of digital propaganda. Employees at Facebook have managed to dismantle predatory and disinformative advertisements on topics from payday loans to which political candidate should get a vote. Googlers have stood firm against shady dealings in military drone research and manufacturing. It is also clear to me that tech's large firms have to get real with themselves. They are now media companies, purveyors of news, curators of information, and, yes, arbiters of truth. They owe a debt to both democracy and the free market, and their allegiance to the latter doesn't mean they can ignore the former.

The Plight of Legacy Social Media

The ongoing attempts by Facebook, YouTube, and Twitter to amend their trend-based algorithms and social network structures to be less vulnerable to bots and false news are chief among the short-term code- or technology-based approaches to tackling the problem at hand. In an ideal world, such fixes would allow users to know when they are being duped. In an even more perfect world, the platforms would have built trending algorithms, especially those associated with the news, so that they could never be duped or manipulated by automated or manipulative accounts. Better yet, these firms would never have launched curatorial "truth-arbitrating" algorithms in the first place.

In our imperfect world, efforts to amend faulty recommendation systems by social media firms have been haphazard as perhaps hundreds of thousands of engineers over the last ten to fifteen years have made mercurial tweaks to these algorithms. It's hard to untangle the web of changes to these algorithms over such long periods of time. It's even harder to figure out who, beyond particular corporations or groups of executives, is responsible for the mistakes that allowed disinformation on social media to challenge the foundations of democratic communication. Again, these companies risk becoming "legacy" social media as quickly as they became "new" media. Ethics and human rights often weren't primary concerns for these companies when their engineers were designing this code. Scaling their product, selling ads, and optimizing user time spent on platforms were more often what drove them. Imagine the complexity of shifting the code behind Facebook, especially now that the algorithms on that platform alone have affected what over two billion people see when they log on. Tool- or software-based solutions are useful in some ways, but ultimately they are infrastructurally complicated and simply not enough on their own.

As mentioned elsewhere, many technological fixes are useless within six months after they're introduced—bugs are found, companies close, user desires shift, the marketplace changes. But more worrying than the brevity of tool-oriented solutions is their tendency to be piecemeal and unidirectional rather than systematic and multifaceted. To use medical terms, they are oriented towards curing dysfunction rather than preventing it. They do not usually involve actual or complete redesign of the digital platforms and tools used by malicious actors to create the problems at hand.

The main problem with these and other efforts by the social media companies and fact-checkers is that, by empowering users, they also shift the burden of solving the problem to them. Why should overwhelmed users be responsible for eliminating bad information rather than multibillion-dollar companies or governments? It's good to be transparent with users, but these efforts can also result in just one more piece of information in the sea of digital data. We are all buried in email, texts, and notifications; it is hard to process yet more information about some random bogus news article from months back that suddenly got fact-checked or marked as Russian propaganda. Furthermore, start-ups can tenaciously and, if nonprofit-driven, fairly selflessly work on software fixes to computational propaganda, but they cannot claim to know the true extent of companies' constant changes to code. As a result, they can never stop racing to keep their products viable. There are very few details on coordinated efforts to deal with the sprawling issue of computational propaganda between the likes of Google and Facebook, let alone between these behemoths and small business innovators or civil society groups.

In the medium and long term, we need better active defense measures against propaganda as well as systematic (and transparent) overhauls of our current social media platforms rather than piecemeal tweaks. We need new social media platforms and new companies. We must move toward more methodical solutions to the problem of computational propaganda. In an article on the topic for the *Guardian*, my colleague Marina Gorbis and I argued for the need for an early warning system for digital deception and propaganda, especially when they are propagated by social bots and automated systems, which are extremely trackable.[4] Noting that scientists track earthquakes and tsunamis by monitoring movements on the ocean floor, we argued, "If we can do this for monitoring our oceans, we can do it for our social media platforms. The principles are the same—aggregating multiple streams of data, making such data transparent, applying the best analytical and computational tools to uncover patterns and detect signals of change." Such approaches cannot only be technical and quantitative. They must also incorporate social knowledge, offline human work, policymaking, and qualitative research.

Using the known ways in which we can protect ourselves against the tide of digital disinformation, we must build informational resilience in our society, a kind of cognitive immunity, and prioritize the values inherent in democracy and human rights in vetting both our data and our technology. We can also strive to protect and empower different social groups using different strategies to facilitate their digital or networked connections online and offline while helping to inoculate them against junk news and fake science by reminding them of their uniqueness and their right to high-quality informational resources. In addition, it's time for governments to get serious about media literacy, news literacy, information literacy, digital literacy—whatever you want to call it. Existing systems for educating people about media and data, as danah boyd and other experts have pointed out, need a lot of work. These approaches, boyd notes, have failed to "take into consideration the cultural context of information consumption that we've created over the last thirty years."[5] We need to build flexible, approachable, and culturally contextual media literacy campaigns for the digital age.

The longest-term solutions to the problems of computational propaganda and the challenges associated with digital political manipulation are analog, offline solutions. We have to invest in society and work to repair damage between groups. We need to figure out ways to allow those with whom we have disagreed, argued, or even fought to redeem themselves. We must accept that we are also imperfect in our own informational habits, try to improve them, and ask for forgiveness where needed. No one shares perfect data all of the time. None of us are always rational. Nationally and internationally, polarization, nationalism, globalization, and extremism have created wide divides among people where once there were only small gaps. These issues can be addressed, but the primary solutions will be social, from investments in our educational systems to amendments to laws or ideas that we may once have thought of as immutable. Technology can help us, but even the most advanced machines and software systems are still tools. They are only as useful as the people, and motives, behind their creation and implementation.

At the end of the day, we cannot just continue to fight technology with more technology. Thinking toward social solutions requires that we accept that polarization, nationalism, globalization, and extremism are the basic problems in our current world, while disinformation and propaganda are the symptoms of those problems. Nevertheless, bad symptoms that intensify can worsen the underlying disease. Small things like disinformation on Twitter can inflame large issues, like polarization, in a rather circular way. But we have to get serious about what we are trying to address, and when and how we are going to do it. We must be systematic in our efforts to fix the problems created by malicious and manipulative uses of social media and other technology. And we need to repair the social bonds that these tactics are so effective at weakening even further.

The Value of Social Research on Technology

In the hundreds, if not thousands, of conversations I've had with current and former tech employees in Silicon Valley and elsewhere about the problems of computational propaganda, qualitative researchers and tech journalists have often been attacked and blamed for misperceptions about digital disinformation. I have regularly been told that people who study experience and gain knowledge through interviews or fieldwork have screwed up the public's understanding of the phenomenon popularly and problematically known as "fake news." Journalists and researchers like me have allegedly blown the rise of disinformation out of proportion, have been generally sensationalistic, and have no access to "real data" on what actually has happened. I may be particularly sensitive to these allegations because I am, at heart, a qualitative researcher. But I am also a pragmatist, and I see the need for multiple types of inquiry in order to fully understand and combat any sociotechnical problem.

Silicon Valley speaks the language of numbers, quantitative analysis, math, computation. If change is to come there, and in other technology hubs around the globe, it is likely that firms will need access to hard numbers. I've been told as much by people I respect and regularly collaborate with, and they say this with complete knowledge that the behavioral data—so precious to quantitative researchers in particular—to which the Googles and Facebooks of the world have access is proprietary. So too are the algorithms that prioritize what users see—whether disinformation or the latest news—and when they see it. It's a challenge to effectively study these things without that data. But I firmly believe that looking at the numbers alone is not enough when coming to conclusions about disinformation, its effects on individuals and society, and potential solutions.

Facebook and other companies are beginning to share (scrubbed and time-bound) data with us on how people acted on their platforms during elections.[6] YouTube has even begun making an effort to stop directing users to conspiracy theory videos.[7] There are well-documented problems with the study of big data, which can be reductive or misleading or fail to map reality effectively. But a quantitative analysis of these platforms is still necessary. It is also true not only that open-source intelligence work on social media is getting stronger but that social media companies are likely to continue to share more data with researchers, even if with painstaking slowness. Most importantly, however, attacks on researchers and tech journalists are as dangerous as they are misinformed.

Qualitative social science has played a huge role in surfacing the many sociopolitical problems caused by unthinking technology design and manipulative uses of the internet. Much of this critical work was pioneered by female researchers. Contemporary scholars, including Lucy Suchmann, Nancy Baym, danah boyd, Annette Markum, Kate Crawford, Gina Neff, Mary Grey, and many others, continue to reveal the systems of power at play in the construction and use of various aspects of the internet and new media systems.

Binary thinking about how we generate understandings of digital disinformation is a problem. We need quantitative work to understand the breadth of inorganic information operations, to draw casual relationships between the spread of junk news content and public opinion, and to determine whether behavioral changes accompany political bot campaigns.

But we need qualitative work to complement these endeavors. As AI Now cofounder Kate Crawford once told me, though much more eloquently, we have to understand the people and groups that build technology in order to understand the impact of digital tools on society. We cannot divorce qualitative understandings of people's norms, values, and beliefs from quantitative statistics on their technology use.

The Evolving Global Problem

There's little doubt that the new frontiers of propaganda require some kind of legal intervention. Foreign governments have used, and continue to use, digital communication tools in attempts to sway the elections of their adversaries. And governments from Russia to China do not reserve these attacks for foreign competition; they also leverage them in efforts to exploit allies and their own citizens. In our concern over this issue, however, we must not overstate the technological sophistication of the attacks that have already occurred. To reiterate, most computational propaganda attacks since 2010 have not been technologically sophisticated. These offensives have been efforts to hack public opinion, not computational infrastructure.

Computational propaganda campaigns, from Russian manipulation of the US election in 2016 to Syrian government attempts to quash online dissent during that country's revolution, have used social media technologies to do exactly what they were designed to do: amplify information, communicate about social life, and generate trends. Those who launch them have simply used platforms like Facebook and Twitter to control rather than liberate—clearly to the shock of the social media companies, which should have had enough foresight to see that powerful political actors, and even regular people, would try to use their platforms to repress at the same time others were using them to democratize.

As I've said earlier, social bots have played a role on Twitter since the site launched, to instantly post not only the latest news stories or banal advertisements but also deluges of conspiracy and propaganda. Warnings upon warnings were given to Silicon Valley companies that their technology was being used by the powerful to manipulate the weak: during the Arab Spring in 2011, the Mexican election in 2012, the Boston Marathon in 2013, and the Turkish election in 2014, and in numerous other situations where people used social media to spread dangerous rumors, disinformation, and political attacks. Most of the attacks, whether driven by bots, humans, or cyborgs, were fairly simple. They didn't use artificial intelligence, machine learning, or deep learning, nor did they involve deepfakes or humanlike technology.

But the era of smarter technology will be upon us soon. These deceptive campaigns will grow more powerful, just as email scams have graduated from free-associative spam messages and Nigerian prince scams to sophisticated phishing attacks. As the more basic manipulative uses of social media—for instance, inorganic armies of accounts manufacturing political trends—become known, propagandists will have to get more clever. In general, software of all kinds is continuously advancing. Machine learning and deep learning, once more theoretical than actual, are becoming operational and cheaper to use in commonplace settings. The increasing availability of artificial intelligence tools has implications for written, visual, auditory, and tactile communication online. Social bots are becoming more interactive, deepfakes are becoming more convincing, and artificial voices, from Siri to Cortana, now sound more humanlike.

With these innovations, and alongside the uptick in the general understanding of digital disinformation, the black-hat PR firms, crooked political consultants, and a slew of other groups that use computational propaganda are altering their tactics. They are changing how they launch their operations on the legacy social media platforms, making their people act more like bots, and their bots more like people, in efforts to confuse the algorithms built to track the inorganic spread of content. These groups are seeding and fertilizing bogus news stories among groups on other platforms—such as WhatsApp and Telegram—in order to coerce and confuse voters. Targeting those they see as particularly vulnerable, the young and the old, in new places and in new ways, they are sowing junk science on TikTok and stoking fear on Instagram. And though, in the last five years or so, internet propagandists have targeted right-wing voters in Brazil, France, India, South Africa, the United States, and the United Kingdom, this too is changing. My own research into the manipulation of social and issue-focused groups during the 2018 US midterms revealed that leftists are increasingly the targets of such campaigns.

In the 2020 US election, it is very likely that the Russian government, for instance, will focus its attacks against the Democrats rather than the Republicans and do so by targeting existing divides—for example, the split between the party's centrist wing and its democratic socialist bloc. Whereas in 2016 likely Republican voters were fed fake stories about Hillary Clinton being corrupt and dishonest, the 2020 electorate may get stories that poison them against centrist candidates like Joe Biden, along with fearmongering stories about candidates like Elizabeth Warren wanting to destroy the stock market. Such stories are particularly likely to target whoever emerges as the front-runner in early 2020, perhaps diverting votes to the second- or third-place candidate or an independent like Howard Schultz.

An alternative strategy would also splinter the left's vote. If particular subsections of the US left are made to believe that the candidacy of a far-left contender was stolen by the mainstream Democratic Party, then they are as likely to not vote at all as they are to vote for the party nominee. During my work on the 2016 election, I saw a great deal of evidence for this type of activity on both the right and left. The data that Facebook shared about Russian manipulation on that platform backs this up. The Russians built manipulative Black Lives Matter and Blue Lives Matter pages, created pro-Muslim and pro-Christian groups, and let them expand via growth from real users. The goal was to divide and conquer as much as it was to dupe and convince.

But it wasn't just the Russians who successfully used social media to manipulate public opinion in 2016. And it will not just be foreign governments that use new technology for such purposes in years to come. Political campaigns will also make use of new technology. In the twelve months preceding the 2016 election, the Trump campaign had spent more on social media than any other candidate, including Clinton. Trump pointed to metrics like follower counts and online surveys as proof that he was winning. I would shake my head at these moments, knowing that these numbers were artificially inflated—millions of those followers were fake, after all. But he was right. Regardless of how bogus the traffic was, it did something more important. It created a bandwagon effect among actual voters and legitimized fringe views that turned out to be supported by a lot of people. Consequently, whatever was being shared had to be taken more seriously by journalists, and their coverage then broadcast those stories—some of them fake, it would turn out—even more widely.

On top of that, the wide array of Republican digital political consultants I interacted with were much savvier than their competition. They were far more willing to use and experiment with the newest tools—some ethically, some not so much—for controlling conversation online. Bots were only part of the strategy. These people were into memetics (information patterns), live-streamed video of the candidate (which showed candor, they said), dark ads (ads that are shown only to certain people on social media, based on their interests), and—above all—false news (which painted Clinton as a despot).

While all this was happening, Democrats were relying on the outdated Obama-era system of using online databases to call independent voters and spending hundreds of millions on local TV ads rather than on Facebook content. The Republicans were experimenting with covertly reaching voters using incredibly granular online data supplied directly by Facebook and Google employees—who had permanent desks at Trump's digital operation but not at Clinton's. As scholars Daniel Kreiss and Shannon McGregor put it, when the companies approached Trump to offer their services, he treated them like consultants. When they approached Clinton, she treated them like vendors.[8] Her digital team thought they didn't need help.

This was perhaps the biggest mistake that Clinton's campaign made. Put digital propaganda aside for a moment and consider the fact that Facebook had extensive personal data on more than three-quarters of the American public—and they were using it, perfectly legally, to help Trump's campaign reach undecided voters in swing states, to embolden his core constituency, and to smear the opposition. Political communication had suddenly, and clearly, become a big part of the business model for Facebook, Twitter, and YouTube. Elections happen every year, all over the world. And the companies had ads to sell and data to share. How will VR social media platforms and new media technology companies factor elections into their business models in years to come? What are the priorities of their CEOs? Who will choose to be their customers?

COINTELPRO

The methods used by the Russian government and other groups to spread computational propaganda are, in part, already established tactics from information and propaganda operations of old. The long history of COINTELPRO, a portmanteau derived from COunter INTELligence PROgrams, is relevant to the computational propaganda campaigns of today. COINTELPRO operatives worked to seed dissent within organizations such as the Black Panthers and anti-Vietnam War activist groups, as well as within the American Indian movement and the feminist movement, in order to take them down from the inside. Now online groups have adopted these tactics. Journalist Sam Greenspan discusses, for instance, the way in which COINTELPRO and online organizing tactics are now used in online communities, such as Reddit, to attack safe spaces for marginalized communities.[9]

In another push to innovate in the computational propaganda space, manipulative groups are now beginning to post advertisements containing false news and other divisive political content on peripheral sites. Now that Google and Facebook have begun to regulate political advertisements—those who buy them must now adhere to certain standards, including clear notice of who paid for a particular ad—propagandists are moving to other social media sites and websites for large special interest communities that have little to no regulation of who can advertise and how.

To prevent targeted attacks and defamation campaigns against the most vulnerable, we must create regulations and policies that protect minority classes from online manipulation. New laws should make it more clearly illegal for social media firms to sell advertisements that target these groups with politically charged misinformation or disinformation. Mainstream social media platforms should provide safe spaces online for these groups and facilitate their day-to-day use by protecting and moderating them. Public forums, whether they resemble Facebook group pages, Twitter feeds, or some not yet created digital space, should be more vigorously policed for both hate speech and information operations. It is not acceptable for social media firms or other technology companies to address computational propaganda on a case-by-case basis, responding seriously only to cases that garner serious media attention. The protection of groups with already marginalized voices—and I mean underrepresented religious, ethnic, or social minorities, not groups that consider themselves marginalized but actually work to further the agenda of the status quo (like so-called men's rights activists)—must be foundational to social media and any new socially oriented technology.

Young People and Future Technology

In the course of my work on computational propaganda, I spend a lot of time talking to adults of various ages, and many of them are concerned about technology use by young people, particularly those under age eighteen. Their concern is ironic considering that we now know that it is often older people who are more likely to share false news or disinformation over social media. Although young people may sometimes be vulnerable to online propaganda, many argue that they are less susceptible to digital manipulation because they have grown up online. They are, in other words, "digital natives."

Nevertheless, propagandists are constantly looking for new ways to mold young people. It makes sense that those looking to coerce and deceive would target kids through their phones, computers, and other connected devices, since they are much more comfortable with this technology than older generations. Social media platforms continue to be spaces where manipulative information makes its way to young people, however, and contrary to the belief of some, recent research underscores the fact that they are particularly bad at distinguishing "fake" news from real news.[10] One study tested students aged 12–22 and collected nearly 8,000 responses. "Many assume that because young people are fluent in social media they are equally savvy about what they find there," the report says. "Our work shows the opposite."

So how might future disinformation campaigns be leveraged against young people? What role will be played by our current technologies, as well as by those not yet created, in efforts to either educate or mislead youth?

We know that young people are more inclined than older people to use image-based technology. They are particularly drawn to applications that allow them to share photos, videos, or even instant streams. According to a 2018 report from the Pew Research Center, Instagram, Snapchat, and YouTube were by far the most popular platforms among US teens.[11] Facebook and Twitter, meanwhile, were becoming less popular among young people. As many as 95 percent of those surveyed had regular access to a smartphone, and 45 percent told researchers that they were online "almost constantly." Mobile instant messengers, like Telegram and WhatsApp, are also becoming increasingly popular with young people around the globe. So too are newer versions of older technologies. MeetMe, Omegle, and Yubo, which introduce people to one another in a chatroom-like atmosphere, are all popular with teens. That they offer varying degrees of anonymity can cause obvious issues. Teens are regularly, if not constantly, connected, and they use social media as a primary form of communication.

Many of these applications have built-in parental controls, but others do not. Moreover, many teens use social media in ways that are well beyond the understanding of their parents. As researcher danah boyd says, in reference to the networked life of teens, "it's complicated."[12] The social media expertise of young people should not, however, prevent us from designing new social media tools that are geared toward protecting them as well as other vulnerable populations, such as the elderly, from malicious or manipulative uses of social media. Start-ups like Soap AI are building news and information applications that reach teens on their own level.[13] Neither boring nor hard to use, these applications aid in promoting media literacy and incorporate, for instance, entertainment news and sports alongside information on politics or science.

Ethical Operating Systems

The relatively short history of digital co-optation and control is our foundation for understanding the future. We cannot effectively combat the computational propaganda that comes next without first addressing the flaws in our system that allow it to thrive now. Automation and anonymity on social media and new technology platforms need to be carefully considered. We must ask: Does allowing someone to be masked protect them from tyrants, or does anonymity allow them to be tyrants themselves? There is no clear yes-or-no answer to this question, which must be asked on a case-by-case basis. We should also consider: Does automation make us better able to consume good information or is it just creating noise? Our current knowledge of how computational propaganda is produced and circulated is our best defense against technological malfeasance that is advancing more rapidly than governmental policy and civic advocacy groups can respond. A well-informed citizenry is at the very core of democracy—and may be our strongest defense against digital autocracy.

The Digital Intelligence Lab's Ethical Operating System (Ethical OS) provides one path for working with those who build and use technology to solve the problems presented by computational propaganda. Besides implementing the existing software and policy-oriented solutions that address particular issues like bot-tracking, we must also work with the next generation of software designers and students to make sure that they learn from the mistakes of their predecessors. As my co-creator Jane McGonigal writes:

> *As technologists, it's only natural that we spend most of our time focusing on how our tech will change the world for the better. Which is great. Everyone loves a sunny disposition. But perhaps it's more useful, in some ways, to consider the glass half empty. What if, in addition to fantasizing about how our tech will save the world, we spent some time dreading all the ways it might, possibly, perhaps, just maybe, screw everything up? No one can predict exactly what tomorrow will bring (though somewhere in the tech world, someone is no doubt working on it). So until we get that crystal ball app, the best we can hope to do is anticipate the long-term social impact and unexpected uses of the tech we create today.*[14]

All of us, but especially technologists, need to foresee the potential damage caused by the tools we create before we make them, not long afterwards.

Our team outlines eight risk zones in the Ethical OS: truth, disinformation, and propaganda; addiction and the dopamine economy; economic and asset inequality; machine ethics and algorithmic biases; the surveillance state; data control and monetization; implicit trust and user understanding; and hateful and criminal actors. If we are to effectively address the problems at hand, all of us, including current and future technologists and policymakers, must be thoroughly educated in each risk zone. Technology companies and the firms in their orbit should require all employees to pass a course on potential misuses of technology. Some have even convincingly argued that software engineers should have to take a Hippocratic oath to, before all else, do no harm.[15]

It is not a given that tomorrow's technology creators, building tools enhanced by artificial intelligence and quantum computing, will prioritize features like anonymity and automation on the front end of social networking. Identity-based metrics could be the next means of curbing disinformation. We must think hard about how we institute identity verification requirements into our tools. There is still a place, as I've said earlier, for anonymity in our digital world. Some people rely on it to survive. With new tools, including the blockchain but also other novel ways of checking on provenance and identity, it is becoming possible to more effectively ID users as real or nonmalicious while still preserving their private information. This said, we cannot maintain blind allegiance to the outmoded ideal of an open internet and the immutable need for anonymity no matter what. Anonymity is a knife that cuts both ways. We can build systems that use this feature to protect the weak, while also constructing tools to prevent malicious actors from using anonymity to attack or manipulate people.

Many social media and technology companies have set out new guidelines for adhering to values like transparency in order to maintain user trust. The worthwhile idea behind transparency is that the more information you give people about a technological system, even about its failures, the better off everyone will be. However, transparency alone is not enough. Facebook and Google cannot simply overwhelm users with data about a given issue or the code behind a particular algorithmic decision. They should not prioritize transparency only when something goes wrong, giving people information after the fact. They must build transparency into their products and their company ethos. More importantly, they should prioritize the translation of information that is often undecipherable to regular users. These firms, and others, must work with third-party researchers and experts to decipher information that is crucial to user experience, as well as information related to users' social and political lives. Impartial organizations that are not in the pocket of technology firms must be invested with the task of auditing code and algorithms for discrimination, faulty decision-making, and other problems.

Fixing the Breakdown of Reality

Where is there hope?

In the United States, one-quarter of adults are online in some way, shape, or form nearly all the time.[16] Internet penetration rates continue to rise around the globe, and there is still a lot more growth to come. In India only 22 percent of a population of 1.3 billion are connected, but the rate continues to grow.[17] Billions more people are slated to go online, and the way they experience the internet will differ markedly from the experiences of those who began using it in the mid-1990s. Is the flow of technology so great that we cannot stem the stream of manipulative uses? Are things changing so rapidly that we can't work to control the negative effects of technology use? I do not think so. I believe we can still redesign old technology systems, build new products grounded in ethical use, and address the sociopolitical issues at hand. But we need to do so now.

Old systems, including social media sites governed by opaque algorithms that can be gamed to spread disinformation, need an overhaul. This has to happen in a coordinated fashion. Companies like Facebook and Google cannot continue to work alone, black-boxing their technology and holding back facts to save face. We need new policy to govern and protect the online sphere, especially legislation that prevents the misuse of the internet during elections and major events. Countries around the globe, from Germany to Brazil, already have laws on the books that prevent certain misuses or other problematic aspects of internet technology and data gathering. But new laws need the input of people who understand technology, people who understand policy, and people who understand society. We need to invest in more public interest technologists who can help to generate sensible laws to govern information flows online and potential misuses of future technology. Governments must work together and accept that computational propaganda is, by its very nature, a transnational problem as much as a multi-platform issue.

I do not believe that we have to do away wholesale with anonymity and automation, and I am not sure we can. Both have purposes that benefit equality, privacy, and free speech and both are arguably infrastructural to some parts of the internet. Recent proposals for laws that get rid of all bots on social media or that kick off all politically inclined bots need serious work and attention to nuance. When thinking about how to govern potentially problematic uses of emerging tools, from XR to voice systems and beyond, we must strive to protect both free speech and privacy, but we must also not allow the proliferation of hate speech, harassment, voter disenfranchisement, harmful health care disinformation, or the gaming of systems in order to prioritize one person's or group's ideas over another. We cannot allow future social media networks, whether on our laptops or on VR headsets, to become cesspools of false news. We can protect innovation while also putting a stop to laissez-faire laws that prioritize money and progress over all else.

People are already beginning to address the problems presented by computational propaganda. We at the Digital Intelligence Lab have presented Ethical OS, for instance, to state and federal lawmakers in the US, tech companies, college students, civil society groups, and venture capitalists. It has been taught in Stanford's design school and its "Introduction to Computer Science" course, and other universities are starting to use it as well. Some of the big technology firms in the San

Francisco Bay Area and elsewhere are using it to train employees. If we can get people to bake human goodness, rather than deceit and growth, into our tools, then we can prevent future issues.

While new laws are being passed and new ethical tools are being built, the technology firms behind the world's social media can address the problems of computational propaganda. Many of these organizations have already begun taking steps to curb the spread of false news, politically motivated trolling, and junk science. They are redesigning portions of their organizations so that they do not share manipulative information or support . Google employees protested against participating in "Project Maven," an internal military drone project, and the company backed out of the deal.[18] Workers at both Facebook and Google spoke out against predatory payday loan advertisements, and the organizations stopped selling them.[19] These companies, and others, have retooled their algorithms in attempts to stop the flow of disinformation.[20] But there is still a lot of work to do.

The Googles and Facebooks out there cannot just address these issues in a piecemeal fashion. To build up democracy and human rights they must take firm stances against manipulative practices as well as computational propaganda in all its forms. At present, most social media firms have three-part business models that prioritize getting users to pay attention, collecting personal data, and creating opaque algorithms to drive content creation and ad targeting. These priorities have to change. The companies cannot keep preying upon their users—upon their happiness, health, and well-being—to make money. Now used by billions, they bear responsibility for modifying and changing the information we see based on their own ideals or ideas. They must alter their business models to allow democracy—not just the bottom line—to succeed.

As we work to unravel the mess made by moving too fast and breaking too many things, many have been advocating antitrust solutions. Proponents argue that organizations like Google and Apple have grown too large and control too much on too many levels—in other words, that they are monopolies and must be broken up. Although I certainly agree that too few technology companies now have too much power in the online world, I do not think the antitrust solution will solve all—or even most—of the problems associated with the current crisis.

Breaking apart the companies, while perhaps necessary for other reasons, will not necessarily help those working to understand computational propaganda. The companies and their products are already disorganized and complex, and they are only just beginning to effectively address parts of the problem. If they are broken up prematurely, or without care, we may have to start all over in the quest to construct a democratic information system. Before we dissolve the companies that produced the tools used to game the truth, we must hold them accountable for helping to fix the issue. It is not enough to simply call for Facebook to be disbanded or for the Google monopoly to be broken up. These companies employ some of the best and brightest minds the world has to offer—and many of these people want to help fix the problems at hand. Furthermore, these companies have the hardware and other assets—including unfathomably deep pockets—that will be crucial to getting a handle on computational propaganda.

While these technology firms work to deal with the myriad problems brought about by manipulative uses of their tools, they are also tasked with figuring out who can effectively moderate, monitor, and enforce new rules and standards. Many companies are turning to outside entities, to contractors, to manage these tasks. I'm very wary of allowing massive tech firms to hire external organizations with little understanding of computational propaganda to manage the mess the firms created in the first place. More than this, we now know from recent revelations about the people who moderate content on these sites that they are underpaid, undertrained, and exposed to horrendous imagery and content without due consideration to potential effects.[21] In many cases, people in India or other countries are working to decipher and moderate content in the United

States. Not only are the cultural constraints a burden and a challenge for these workers, but so too is the fact that they are paid much less than their US counterparts.

We should not accept it as a given that this oversight work must be outsourced or automated. Regardless of where the work is done, companies must pay fair wages for this particularly traumatic task. They must also ensure that moderators and those working to stop computational propaganda are rigorously trained in the phenomena they are working to prevent and provide them with equitable employment status and access to quality psychological care. We need new and improved ways to moderate political content, harassment, and trolling and new tools for tracking misinformation as it spreads. We also need new systems for tracking information operations before they take hold. Technology firms must lead the charge in creating these tools and policies, and governments must pass laws that regulate and uphold them.

Online platforms have a great deal of work to do in protecting our privacy, identifying and exterminating malicious automation, and stopping the flow of online propaganda. These entities must also work to prevent the misuse of future technology. But governments have a lot to answer for as well. It is an absolute outrage that so few laws have been passed to address political communication online in the United States and many other countries. The US legal system is in the dark ages when it comes to dealing with problems online in general, let alone with more specific issues like social media. Lawmakers must work to protect the hundreds of millions of Americans who have already been deceived for political purposes during past elections. The US government is responsible for regulating Silicon Valley, not rolling over while it becomes the playground for two or three monopolies. New regulation informed by the expertise of people who actually understand how technology works is a necessity for preventing the degradation of the internet and the misuse of future technology.

In a recent paper, Ann Ravel, Hamsini Sridharan, and I outline a slew of sensible and simple policies that can be immediately enacted to curb the effects of digital deception.[22] We also propose a number of systemic changes that could be instituted to prevent future misuse of digital platforms. Building on an earlier report from Sridharan and Ravel, we recommend the following US-oriented policy actions for immediately illuminating problems at the intersection of computational propaganda and campaign finance:

1. Pass the Honest Ads Act mandating that major technology platforms publish political ad files in order to increase transparency about money in digital politics and dark advertising. Ensure that the data provided is standardized across platforms and provides the necessary level of detail regarding target audiences and ad spends. Records should remain publicly available for several years to facilitate enforcement.

2. Expand the definition of "electioneering communications" to include online ads and extend the window for communications to qualify as "electioneering." Electioneering communications are ads on hotbutton issues that air near an election and reference a candidate, but do not explicitly advocate for or against that candidate. Currently, online ads are exempted from the disclosure rules for this type of advertising, which apply to TV, radio, and print. Online ads that satisfy the definition of electioneering communications ought to be regulated. Moreover, with political ads running earlier and earlier each election cycle, it is important to extend the window of time that electioneering regulations apply for online advertising.

3. Increase disclosure requirements for paid issue ads, which frequently implicitly support or oppose candidates and are intended to motivate political action, but receive little oversight. This is one area where the

government is hampered by court interpretations of free speech, but where technology companies could successfully intervene with civil society guidance.

4. Increase transparency for the full digital advertising ecosystem by requiring all political committees to disclose spending by subvendors to the FEC. Right now, committees must report payments made to consultants and vendors, but aren't required to disclose payments made by those consultants and vendors for purchases of ads, ad production, voter data, or other items; as a result, much digital activity goes unreported. California requires political committees to report all payments of $500 or more made by vendors and consultants on their behalf. Similar rules should be adopted at the federal level.

5. Adapt on-ad "paid for by" disclaimer regulations to apply to digital advertising. Digital political ads should be clearly labeled as promoted content and marked with the name of whoever purchased them. They must contain a one-step mechanism, such as a hyperlink or pop-up, for accessing more detailed disclaimer information, including explicit information about who is being targeted by the ad. There should be no exceptions to this rule based on the size of ads; unlike with pens or buttons, technology companies can adapt the size of digital ads to meet legal requirements.

6. Create an independent authority empowered to investigate the flows of funding in digital political activity. This equivalent to the Financial Industry Regulatory Authority would be charged with following the money to its true sources, making it easier for the FEC to identify violations and illegal activity and enforce penalties.

Each of these efforts would result in a clearer and less deceptive digital space. By passing legislation to require more transparency in social media advertising, more thorough investigation of digital political activity, and, simply put, more accountability in how politics gets done online, we will build a more democratic online world.

But this is only one area among many where we need to pass new laws and policies to regulate the social media sphere. Solutions are needed for problems posed by data usage and privacy, automation and fake accounts, platform liability, and multisector infrastructure. These solutions—both their inception and their implementation—require more effective global cooperation around the issue of digital deception, better research and development on this topic, and clearer media and civic education efforts. There are many different strategies that could be tried—including antitrust actions against the tech industry and some kind of global governing board that oversees communication—especially political communication—on digital platforms. It's clear that computational propaganda has numerous sources, and each needs to be dealt with in thoughtful, bespoke ways if we want to have a healthy democratic ecosystem.

No one organization or government can attempt to take on computational propaganda or global attempts to game reality on its own. Nor can one policy or law, or one software product or novel tool, stop the global spread of misinformation and false news. It will take all major social media and technology firms, and many minor ones, working in concert to stop the degradation of the truth. If they are to push back against the notion of "alternative facts," then policymakers in the United States must begin to act. They should follow their colleagues in other countries around the globe, from Japan to Sweden, in championing legislation that works to build a more vibrant, equal, and free online sphere while instituting policy that mitigates manipulation of this space. Federal politicians in the United States can also look to state lawmakers in California, New York, Washington, and elsewhere to garner ideas on legal solutions.

The ultimate power in pushing these groups to make the changes, and avoid the perils, outlined in this book lies with you and me. The combined voices of the people who use today's social media and of those who will use the social media equivalents of tomorrow can generate persuasive arguments about why we must change the way we build our technology and how we are going to do it.

Re-creating Democracy in Our Generation

Many, myself included, have said that the rise of computational propaganda and other misuses of technology have seriously damaged democracy. The social media companies have even discussed the issue in public. "Facebook was originally designed to connect friends and family—and it has excelled at that," wrote Samidh Chakrabarti, Facebook's civic engagement product manager, in a post to the company's blog. "But as unprecedented numbers of people channel their political energy through this medium, it's being used in unforeseen ways with social repercussions that were never anticipated."[23] Chakrabarti elaborated and even took a portion of the blame on behalf of the social media firm: "In 2016, we at Facebook were far too slow to recognize how bad actors were abusing our platform. We're working diligently to neutralize these risks now."

As I mentioned earlier in this book, social media platforms are not the first media tools to undermine the tenets of democracy. Nor will they be the last. In our world there will always be people working to build and maintain power by exerting control over others. They will use whatever means are necessary and available to achieve these goals, including new technologies. Because social media platforms are immensely popular, used by billions of people, manipulative individuals and entities now employ them to further their own goals among as many people as possible. But the latest technologies aren't just useful because they provide access to lots of people—they are also potent because they are automated, anonymous, quick, and increasingly intelligent in their own right. As new forms of video, virtual reality, augmented reality, voice emulation systems, and technology built in the human image arise, the powers that be will attempt to use them as well to bend the truth in their favor.

In some ways, this prospect is simply part of the human story. As we advance, both socially and technologically, we outgrow previous modes of governance and thinking. We progress. We also, at times, regress. And in every generation we are forced to grapple with the task of rebuilding democracy. This has been a perennial struggle since the fall of ancient Rome, and at times democracy has fallen by the wayside for generations. We must not allow that to happen in our time. We can address the social and technological problems at hand in order to create a stronger world, one with systems built to prioritize human rights and human freedoms. As US park ranger and author Betty Reid Soskin said, "democracy will never be fixed" and "we all have the responsibility to form that more perfect union."[24] What we are experiencing is an ongoing confrontation, and it is up to all of us to respond.

When I first began doing work on politics and technology, I genuinely never thought that I would experience a moment like the 2016 US election—a moment when tools we once thought were so promising for democracy and the open flow of information would be co-opted by governments and a variety of other parties to spread lies, harassment, and spin. Nor did I think I'd see social media platforms being used in the coercive and problematic ways they were in electoral contests and pivotal moments in other countries all over the world. I did not start studying computational propaganda out of fear; in fact, the opposite is closer to the truth. I was just as excited by the potential of our new media tools for advancing freedoms as I was concerned about the possibilities of misuse or exploitation.

I have since come to believe that social media tools—and indeed, all the new technology just coming out and on the horizon—can still be tremendously useful in advancing the best parts of humanity. Although, in the past few years, we have lost our way, we can still find true north again. Social media companies have already begun to take responsibility for remedying the issues with the tools and devices they constructed. Policymakers around the world are passing laws to curb political manipulation, targeted harassment, and other problems that have arisen on and via new media. Journalists continue to seek out and report on the truth, based on facts and evidence. We can all, as regular people but also as a global collective, take a lesson from the Quakers and "speak truth to power."[25]

The work is not yet done. In fact, it will never be done. We will always have to fight to maintain a version of democracy that prioritizes the rights and voices of all people. There will always be inequity, and power will often become concentrated in the hands of too few, but we can use the same technology used to control us to speak out against that control. Together, we can demand that our technology be designed with human rights in mind. We can insist that society function with democracy in mind.

Glossary

API: "Application programming interface". A digital interface built to facilitate communication between a client and a server that works to simplify software construction for clients. Through Twitter's various APIs, for instance, people can build and launch various software programs, from social bots to personalized customer experience apps.

Astroturf: Originally, a term used to refer to synthetic grass. In politics, "Astroturf" refers to activity built to look like real citizen-led ("grassroots") campaigning. These undercover operations are often sponsored by corporations, public relations groups, or advertising companies working on behalf of particular politicians or causes.

Augmented Reality (AR): A hybrid interactive media experience in which people use digital devices to display computer-generated content upon a given real-world environment. Pokémon Go is a quintessential example.

Blockchain: A computer-based record-keeping system built to catalogue transactions online. Blockchain is the foundational technology behind cryptocurrency. It is prized because it was constructed to be unalterable, encrypted and easy to validate.

Bots: Broadly defined as an online automated software program, bots can be made to communicate with both computer systems and human users.

- Botnets: Groups of automated software programs. This term can also refer to a collection of socially oriented bots used to communicate over social media. It can also refer to a group of personal computers that have been infiltrated by a software virus (generally unbeknownst to the device owner) that are then co-opted for various purposes including distributed-denial of service attacks (DDoS—see below) and sending spam messages.
- Chatbots: An automated software program that is designed

to communicate with human users and other bot programs over a chat interface. Chatbots have been used online since the early days of the internet and its precursors.

• Social bots/social media bots: Social bots are a type of chatbot used specifically to engage in communication with both humans and other computer programs over social media sites like Facebook and Twitter. Social bots can be used to interact with human users by actually producing written text or posting content but they can also be constructed to "like" content or reproduce it in an effort to manipulate the code that generates social media trends.

Computational Propaganda: The use of automation and algorithms in attempts to manipulate public opinion online. The term was coined by researchers on the University of Washington's Computational Propaganda Project, which has since relocated to the Oxford Internet Institute at the University of Oxford.

Distributed Denial of Service (DDoS) Attacks: An online attack in which an adversary leverages a botnet, or group of computers, to shut-down a network or particular computer. DDoS attacks have been used, for instance, to barrage government websites with so much traffic that they shut down or become unavailable.

Deepfakes: An amalgamation of the words "deep learning" and "fake". These fake videos are altered using artificial intelligence, specifically deep learning software (see below). Deepfakes use a technology known as a generative adversarial network (GAN) to manipulate videos of a subject using other extant imagery or video of that subject.

Deep learning: A branch of machine learning, which is itself a branch of artificial intelligence. Deep learning deploys autonomous networks to learn from sets of information or data that have not previously been labelled or otherwise lacks structure (whereas machine learning requires structured data). Deep learning networks make use of artificial neural networks (ANN), which are based on communication mechanisms in biological brains.

DNC: Democratic National Committee. The official governing body of the U.S. Democratic Party.

Extended Reality (XR): A catch-all term used to refer to the various interactive and simulation-oriented digital tools which both include virtual reality (VR) and augmented reality (AR).

Generative adversarial network (GAN): A type of machine-learning modelling tool used to generate new data that look like initial, or training, data. This systematic technology is particularly useful in generating realistic-seeming images or videos of a person that are, in fact, manipulated using other extant images of that person.

Machine Learning: A branch of artificial intelligence tools that work to autonomously catalogue structured data sets by following patterns and other indicators.

Mixed Reality (MR): Another term, similar to XR, used to refer to the coalescence of the "real" and "digital" worlds. In a mixed-reality world, both computer-generated and physical information can be displayed via a given medium or interface.

PAC (or super PAC): Political action committee, a tax-exempt political group in the U.S. generally created to raise money for or against particular candidates and causes. Super PACS are a type of PAC that cannot make direct investments in a candidate or cause, but can spend unlimited funds in activities "related" to a given campaign. Super PACs, unlike PACs, can raise money from a variety of sources without legal restrictions on the size of a given donation.

RNC: Republican National Committee. The official governing body of the U.S. Republican Party.

Shadow banning: An act in which a platform blocks or bans a given user or group of users without them having knowledge they have been blocked or banned. The term is often associated with or used by conspiracy groups who believe they have been unfairly silenced by a given platform.

Style transfer: A type of software created to alter or manipulate images or videos. Style transfer allows a given image to be reformed in the fashion of another image. One could, for instance, use this technology to alter a random a portrait or landscape photo in the style of artists like Kahlo or Monet.

Technology incubator: An organization that works with "start-up" concepts or new technology businesses in early-stage development. Incubators often provide resources in the form of space and various forms of intellectual and collaborative support.

Virtual Reality (VR): The most well-known form of XR. VR is a simulation technology in which people wear a wide-variety of digital goggles, gloves, suits and other "connected" clothing via which they experience a digitally generated environment.

Endnotes

Author's Note

1 United Nations, "Global Issues: Human Rights," August 30, 2016, http://www.un.org/en/sections/issues-depth/human-rights/.

Chapter One

1 Pia Ranada, "Duterte Says Online Defenders, Trolls Hired Only during Campaign," *Rappler*, July 25, 2017, http://www.rappler.com//nation/176615-duterte-online-defenders-trolls-hired-campaign.

2 Samantha Bradshaw and Philip N. Howard, "Troops, Trolls, and Troublemakers: A Global Inventory of Organized Social Media Manipulation," Computational Propaganda Project Working Paper 2017.12, University of Oxford, July 17, 2017, https://comprop.oii.ox.ac.uk/research/troops-trolls-and-trouble-makers-a-global-inventory-of-organized-social-media-manipulation/.

3 Maria A. Ressa, "Propaganda War: Weaponizing the Internet," *Rappler*, October 3, 2016, http://www.rappler.com//nation/148007-propaganda-war-weaponizing-internet.

4 Pia Ranada, "Duterte Says Online Defenders, Trolls Hired Only during Campaign" (video), *Rappler*, July 24, 2017, https://www.youtube.com/watch?time_continue=56&v=9WnRwiuMc68.

5 US Department of Labor, Bureau of Labor Statistics, "CPI Inflation Calculator," https://www.bls.gov/data/inflation_calculator.htm (accessed April 26, 2019).

6 Ng Yik-tung, Sing Man, and Xi Wang, "China's Ruling Party Trials Virtual Reality Tests of Members' Loyalty," *Radio Free Asia*, May 8, 2018, https://www.rfa.org/english/news/china/tests-05082018111042.html.

7 P. W. Singer and Emerson T. Brooking, *LikeWar: The Weaponization of Social Media* (Boston: Houghton Mifflin Harcourt/Eamon Dolan, 2018).

8 John Haltiwanger, "Jamal Khashoggi Was Barred from Writing in Saudi Arabia after He Criticized Trump, Then Left His Native Country," *Business Insider*, November 20, 2018, https://www.businessinsider.com/why-jamal-khashoggi-left-saudi-arabia-writing-ban-2018-10.

9 Katie Benner, Mark Mazzetti, Mark Hubbard, and Ben Isaac, "Saudis' Image Makers: A Troll Army and a Twitter Insider," *New York Times*, November 1, 2018, https://www.nytimes.com/2018/10/20/us/politics/saudi-image-campaign-twitter.html.

10 Alex Stamos, "An Update on Information Operations on Facebook," *Facebook Newsroom*, September 6, 2017, https://newsroom.fb.com/news/2017/09/information-operations-update/; "The Rise and Rise of Fake News," *BBC News*, November 6, 2016, https://www.bbc.com/news/blogs-trending-37846860.

11 Garth S. Jowett and Victoria J. O'Donnell, *Propaganda and Persuasion*, 5th ed. (Thousand Oaks, CA: Sage Publications, 2011).

12 Jan N. Bremmer, "Myth as Propaganda: Athens and Sparta," *Zeitschrift Für Papyrologie und Epigraphik* 117 (1997): 9–17.

13 Philip N. Howard, "Social Media and the New Cold War," *Reuters*, August 1, 2012, http://blogs.reuters.com/great-debate/2012/08/01/social-media-and-the-new-cold-war/.

14 Jowett and O'Donnell, *Propaganda and Persuasion*.

15 Kate Starbird, Jim Maddock, Mania Orand, Peg Achterman, and Robert M. Mason, "Rumors, False Flags, and Digital Vigilantes: Misinformation on Twitter after the 2013 Boston Marathon Bombing," *iConference 2014 Proceedings* (March 1, 2014): 654–662, doi:10.9776/14308.

16 Anas Qtiesh, "Spam Bots Flooding Twitter to Drown Info about #Syria Protests," Global Voices Advocacy, April 18, 2011, http://advocacy.globalvoicesonline.org/2011/04/18/spam-bots-flooding-twitter-to-drown-info-about-syria-protests/; Jack Stubbs, Katie Paul, and Tuqa Khalid, "Fake News Network vs. Bots: The Online War around Khashoggi Killing," *Reuters*, November 1, 2018, https://www.reuters.com/article/us-saudi-khashoggi-disinformation-idUSKCN1N63QF.

Chapter Two

1 For a detailed map of the event, see Jon Ostrower, Alexander Kolyandr, Margaret Coker, and Paul Sonne, "Map of a Tragedy: How Malaysia Airlines Flight 17 Came Apart over Ukraine," *Wall Street Journal*, n.d., http://graphics.wsj.com/mh17-crash-map.

2 Matt Viser, "Conservative Group Used Tweet Strategy against Coakley," *Boston Globe*, May 4, 2010, http://archive.boston.com/news/nation/articles/2010/05/04/conservative_group_used_tweet_strategy_against_coakley/.

3 Panagiotis Takis Metaxas, Eni Mustafaraj, and Daniel Gayo-Avello, "How (Not) to Predict Elections," in 2011 IEEE Third International Conference on Privacy, Security, Risk, and Trust (PASSAT) and 2011 IEEE Third International Conference on Social Computing (SocialCom), October 9–11, 2011, 165–171, http://ieeexplore.ieee.org/xpls/abs_all.jsp?arnumber=6113109.

4 Edgar B. Herwick III, "How Roman Catholics Conquered Massachusetts: The Inside Story," *WGBH News*, April 10, 2015, https://www.wgbh.org/news/post/how-roman-catholics-conquered-massachusetts-inside-story.

5 Joan Frawley Desmond, "Mass. Catholics Bank on Scott Brown," *National Catholic Register*, January 20, 2010, http://www.ncregister.com/daily-news/massachusetts_catholics_bank_on_scott_brown; Kathryn Jean Lopez, "It's a Good Thing for Martha Coakley That There Are No Catholics in Massachusetts," *National Review*, January 14, 2010, https://www.nationalreview.com/corner/its-good-thing-martha-coakley-there-are-no-catholics-massachusetts-kathryn-jean-lopez/.

6 For more information on manufactured controversies in science, see Leah Ceccarelli, "Manufactured Scientific Controversy: Science, Rhetoric, and Public Debate," *Rhetoric and Public Affairs* 14, no. 2 (2011): 195–228.

7 Malcolm Gladwell, "Small Change: Why the Revolution Won't Be Tweeted," *New Yorker*, September 27, 2010.

8 Evgeny Morozov, "The Brave New World of Slacktivism," *Foreign Policy* 19, no. 5 (2009).

9 W. Lance Bennett and Alexandra Segerberg, *The Logic of Connective Action: Digital Media and the Personalization of Contentious Politics* (New York: Cambridge University Press, 2013).

10 Philip N. Howard, *The Digital Origins of Dictatorship and Democracy: Information Technology and Political Islam* (New York: Oxford University Press, 2010).

11 Samuel Woolley and Philip N. Howard, "Social Media, Revolution, and the Rise of the Political Bot," in *Routledge Handbook of Media, Conflict, and Security* (London: Taylor and Francis, 2016).

12 Samuel Woolley, "Bots Aren't Just Service Tools—They're a Whole New Form of Media," *Quartz*, April 10, 2017, https://qz.com/954255/bots-are-the-newest-form-of-new-media/.

13 Samuel Woolley, Samantha Shorey, and Philip Howard, "The Bot Proxy: Designing Automated Self Expression," in *A Networked Self and Platforms, Stories, Connections* (New York: Taylor and Francis, 2018).

14 Nour Al Ali and Selina Wang, "Russian-Linked Bots Used US Startups to Meddle in Elections," *Bloomberg*, October 19, 2018, https://www.bloomberg.com/news/articles/2018-10-19/russia-linked-bots-used-u-s-startups-to-meddle-in-election.

15 Washington Post staff, "Full Transcript: Sally Yates and James Clapper Testify on Russian Election Interference," *Washington Post*, May 8, 2017, https://www.washingtonpost.com/news/post-politics/wp/2017/05/08/full-transcript-sally-yates-and-james-clapper-testify-on-russian-election-interference/.

16 Tony Romm, "Trump Met with Twitter CEO Jack Dorsey—and Complained about His Follower Count," *Washington Post*, April 24, 2019, https://www.washingtonpost.com/technology/2019/04/23/trump-meets-with-twitter-ceo-jack-dorsey-white-house/?utm_term=.f0e95cd37947.

17 Norah Abokhodair, Daisy Yoo, and David W. McDonald, "Dissecting a Social Botnet: Growth, Content, and Influence in Twitter," presented at the Eighteenth ACM Conference on Computer Supported Cooperative Work and Social Computing, Vancouver, BC, March 14–18, 2015, https://doi.org/10.1145/2675133.2675208, 839–851.

18 FBI, "Syrian Cyber Hackers Charged: Two from 'Syrian Electronic Army' Added to Cyber's Most Wanted," Federal Bureau of Investigation, March 22, 2016, https://www.fbi.gov/news/stories/two-from-syrian-electronic-army-added-to-cybers-most-wanted.

19 Jose Nava and Guadalupe Correa-Cabrera, "Drug Wars, Social Networks, and the Right to Information: The Rise of Informal Media as the Freedom of Press's Lifeline in Northern Mexico," in *A War That Can't Be Won: Binational Perspectives on the War on Drugs*, ed. Tony Payan et al. (Tucson: University of Arizona Press, 2013), 96–118; "Turkey PM Erdogan Defiant over Twitter Ban," *Al Jazeera*, March 23, 2014, http://www.aljazeera.com/news/middleeast/2014/03/turkey-

pm-erdogan-defiant-over-twitter-ban-2014323164138586620.html; Muhammad Nihal Hussain, Serpil Tokdemir, Nitin Agarwal, and Samer Al-Khateeb, "Analyzing Disinformation and Crowd Manipulation Tactics on YouTube," presented at the 2018 IEEE/ACM International Conference on Advances in Social Networks Analysis and Mining (ASONAM), Barcelona, Spain, August 28, 2018, https://doi.org/10.1109/ASONAM.2018.8508766, 1092–1095.

20 Ariana Tobin, "Facebook Promises to Bar Advertisers from Targeting Ads by Race or Ethnicity. Again.," *ProPublica*, July 25, 2018, https://www.propublica.org/article/facebook-promises-to-bar-advertisers-from-targeting-ads-by-race-or-ethnicity-again.

21 Daniel Kreiss and Shannon C. McGregor, "Technology Firms Shape Political Communication: The Work of Microsoft, Facebook, Twitter, and Google with Campaigns during the 2016 US Presidential Cycle," *Political Communication* 35, no. 2 (April 3, 2018): 155–177, https://doi.org/10.1080/10584609.2017.1364814.

22 Katerina Eva Matsa and Elisa Shearer, "News Use across Social Media Platforms 2018," Pew Research Center, *Journalism & Media*, September 10, 2018, http://www.journalism.org/2018/09/10/news-use-across-social-media-platforms-2018/.

23 Organized Crime and Corruption Reporting Project, "About Us," https://www.occrp.org/en/about-us (accessed February 12, 2019).

24 Jo Ling Kent, "Bots Are Stealing Your Social Media Identity—and Making Money off You," *NBC News*, January 28, 2018, https://www.nbcnews.com/tech/social-media/twitter-bots-are-stealing-social-media-identities-profit-n841951.

25 Soroush Vosoughi, Deb Roy, and Sinan Aral, "The Spread of True and False News Online," *Science* 359, no. 6380 (March 9, 2018): 1146–1151, https://doi.org/10.1126/science.aap9559.

26 Paul Mozur, "A Genocide Incited on Facebook, with Posts from Myanmar's Military," *New York Times*, October 18, 2018, https://www.nytimes.com/2018/10/15/technology/myanmar-facebook-genocide.html.

27 Vindu Goel, Suhasini Raj, and Priyadarshini Ravichandran, "How WhatsApp Leads Mobs to Murder in India," *New York Times*, July 18, 2018, https://www.nytimes.com/interactive/2018/07/18/technology/whatsapp-india-killings.html.

28 "Twitter Dominated by Far-Right and Political Groups Fuelling Anti-Refugee Sentiment during Key Period of the Refugee Crisis," Dublin City University, December 4, 2018, https://www.dcu.ie/news/news/2018/Dec/Twitter-dominated-far-right-and-political-groups-fuelling-anti-refugee-sentiment.

29 Jefferson Graham, "'Crisis Actors' YouTube Video Removed after It Tops 'Trending' Videos," *USA Today*, February 21, 2018, https://www.usatoday.com/story/tech/talkingtech/2018/02/21/crisis-actors-youtube-david-hogg-video-removed-after-tops-trending-video/360107002/.
30 Google News Initiative, "Building a Stronger Future for Journalism," https://newsinitiative.withgoogle.com/.

Chapter Three

1 Eric Lubbers, "There Is No Such Thing as the Denver Guardian, Despite That Facebook Post You Saw," *Denver Post*, November 5, 2016, https://www.denverpost.com/2016/11/05/there-is-no-such-thing-as-the-denver-guardian/.
2 Laura Sydell, "We Tracked Down a Fake-News Creator in the Suburbs. Here's What We Learned," *All Things Considered*, NPR, November 23, 2016, https://www.npr.org/sections/alltechconsidered/2016/11/23/503146770/npr-finds-the-head-of-a-covert-fake-news-operation-in-the-suburbs.
3 Robert Gorwa and Douglas Guilbeault, "Unpacking the Social Media Bot: A Typology to Guide Research and Policy," *Policy and Internet*, August 10, 2018, https://doi.org/10.1002/poi3.184.
4 Gabriella Coleman, "Hacker Politics and Publics," *Public Culture* 23, no. 3(65) (September 21, 2011): 511–516.
5 Fred Turner, *From Counterculture to Cyberculture* (Chicago: University of Chicago Press, 2006).
6 Fred Vogelstein, "Facebook Just Learned the True Cost of Fixing Its Problems," *Wired*, July 25, 2018, https://www.wired.com/story/facebook-just-learned-the-true-cost-of-fixing-its-problems/.
7 Benjamin Goggin, "More than 2,200 People Lost Their Jobs in a Media Landslide so Far This Year," *Business Insider*, February 1, 2019, https://www.businessinsider.com/2019-media-layoffs-job-cuts-at-buzzfeed-huffpost-vice-details-2019-2.
8 John Pavlik, "The Impact of Technology on Journalism," *Journalism Studies* 1, no. 2 (January 1, 2000): 229–237, https://doi.org/10.1080/14616700050028226.
9 Nathalie Maréchal, "Targeted Advertising Is Ruining the Internet and Breaking the World," *Motherboard*, November 16, 2018, https://motherboard.vice.com/en_us/article/xwjden/targeted-advertising-is-ruining-the-internet-and-breaking-the-world.

10 Sydney Jones, "Online Classifieds," Pew Research Center, *Internet & Technology*, May 22, 2009, http://www.pewinternet.org/2009/05/22/online-classifieds/.

11 Leon Neyfakh, "Do Our Brains Pay a Price for GPS?," *Boston Globe*, August 18, 2013, https://www.bostonglobe.com/ideas/2013/08/17/our-brains-pay-price-for-gps/d2Tnvo4hiWjuybid5UhQVO/story.html.

12 Nick Monaco and Carly Nyst, "State-Sponsored Trolling: How Governments Are Deploying Fake News as Part of Broader Harassment Campaigns," Institute for the Future, February 2018, http://www.iftf.org/fileadmin/user_upload/images/DigIntel/IFTF_State_sponsored_trolling_report.pdf; Samuel Woolley and Philip Howard, *Computational Propaganda: Political Parties, Politicians, and Political Manipulation on Social Media* (New York: Oxford University Press, 2018).

13 See Yochai Benkler, *The Wealth of Networks: How Social Production Transforms Markets and Freedom* (New Haven, CT: Yale University Press, 2006); for a fantastic history of the people who built the web, see Turner, *From Counterculture to Cyberculture*.

14 Dennis Prager, "If You Want a Conservative Child," *The Dennis Prager Show*, November 12, 2013, https://www.dennisprager.com/want-conservative-child/.

15 Joseph Bernstein, "Teach Them Right: How PragerU Is Winning the Right-Wing Culture War without Donald Trump," *BuzzFeed News*, March 3, 2018, https://www.buzzfeednews.com/article/josephbernstein/prager-university (accessed April 30, 2019).

16 Monaco and Nyst, "State-Sponsored Trolling."

17 Samuel Woolley and Katie Joseff, "Computational Propaganda, Jewish-Americans, and the 2018 Midterms: The Amplification of Anti-Semitic Harassment Online," Anti-Defamation League, November 2018, https://www.adl.org/resources/reports/computational-propaganda-jewish-americans-and-the-2018-midterms-the-amplification.

18 Jevin West and Carl Bergstrom, "Calling Bullshit: Data Reasoning in a Digital World," https://callingbullshit.org/syllabus.html (accessed January 13, 2019).

19 "What Is 2017's Word of the Year?" *BBC News*, November 2, 2017, https://www.bbc.com/news/uk-41838386.

20 Solon Barocas, Sophie Hood, and Malte Ziewitz offer a great summary and critique of this and related ideas in "Governing Algorithms: A Provocation Piece," March 29, 2013, https://papers.ssrn.com/sol3/papers.cfm?abstract_id=2245322.

21 Michelle Nijhuis, "How to Call BS on Big Data: A Practical Guide," *New Yorker*, June 3, 2017, https://www.newyorker.com/tech/annals-of-technology/how-to-call-bullshit-on-big-data-a-practical-guide.

22 Gallup, "In Depth: Confidence in Institutions," https://news.gallup.com/poll/1597/Confidence-Institutions.aspx (accessed January 14, 2019).

23 Cary Funk, "Mixed Messages about Public Trust in Science," *Issues in Science and Technology* 34, no. 1 (Fall 2017), https://issues.org/real-numbers-mixed-messages-about-public-trust-in-science/.

24 John Harrington, "Trust in Traditional Media Grows – but UK now 'a nation of news avoiders' says Edelman Trust Barometer," *PR Week*, January 22, 2018, https://www.prweek.com/article/1455051/trust-traditional-media-grows-uk-a-nation-news-avoiders-says-edelman-trust-barometer

25 Nic Newman with Richard Fletcher, Antonis Kalogeropoulos, and Rasmus Kleis Nielsen, "Reuters Institute Digital News Report 2019," Reuters Institute and University of Oxford, June 12, 2019, https://reutersinstitute.politics.ox.ac.uk/sites/default/files/2019-06/DNR_2019_FINAL_1.pdf.

26 Uri Friedman, "Trust Is Collapsing in America," *Atlantic*, January 21, 2018, https://www.theatlantic.com/international/archive/2018/01/trust-trump-america-world/550964/.

27 "Freedom in the World 2018: United States Profile," Freedom House, January 5, 2018, https://freedomhouse.org/report/freedom-world/2018/united-states.

28 Tom McCarthy, "Is Donald Trump an Authoritarian? Experts Examine Telltale Signs," *Guardian*, November 18, 2018, https://www.theguardian.com/us-news/2018/nov/18/is-donald-trump-an-authoritarian-experts-examine-telltale-signs.

29 "Freedom in the World 2018: United Kingdom Profile," Freedom House, January 5, 2018, https://freedomhouse.org/report/freedom-world/2018/united-kingdom.

30 Shanthi Kalathil and Taylor C. Boas, *Open Networks, Closed Regimes: The Impact of the Internet on Authoritarian Rule* (Washington, DC: Brookings Institution Press, 2010).

31 Associated Press, "FEC Won't Regulate Most Online Political Activity," *NBC News*, March 27, 2006, http://www.nbcnews.com/id/12034995/ns/technology_and_science-tech_and_gadgets/t/fec-wont-regulate-most-online-political-activity/.

32 Senator J. James Exon (Democrat-NE) and Senator Gorton Slade (Republican-WA), "Communications Decency Act of 1995," S.314, 104th Cong. (1995–1996), https://www.congress.gov/bill/104th-congress/senate-bill/314/cosponsors.

33 Emily Dreyfuss, "German Regulators Just Outlawed Facebook's Whole Ad Business," *Wired*, February 7, 2019, https://www.wired.com/story/germany-facebook-antitrust-ruling/.

34 Carolina Rossini, "New Version of Marco Civil Threatens Freedom of Expression in Brazil," Electronic Frontier Foundation, November 9, 2012, https://www.eff.org/deeplinks/2012/11/brazilian-internet-bill-threatens-freedom-expression; "Europe Bets Its Data Law Will Lead to Tech Supremacy," *Financial Times*, April 30, 2018, https://www.ft.com/content/f77c3b3a-4c44-11e8-97e4-13afc22d86d4.

35 Christopher Paul and Miriam Matthews, "The Russian 'Firehose of Falsehood' Propaganda Model: Why It Might Work and Options to Counter It," Rand Corporation, 2016, https://www.rand.org/pubs/perspectives/PE198.html.

36 Marco della Cava, "Oculus Cost $3B Not $2B, Zuckerberg Says in Trial," *USA Today*, January 17, 2017, https://www.usatoday.com/story/tech/news/2017/01/17/oculus-cost-3-billion-mark-zuckerberg-trial-dallas/96676848/.

37 Andy Kangpan, "Bright Spots in the VR Market," *TechCrunch*, December 2018, http://social.techcrunch.com/2018/12/02/bright-spots-in-the-vr-market/.

38 Jon Bruner, "Why 2016 Is Shaping Up to Be the Year of the Bot," *O'Reilly Media*, June 15, 2016, https://www.oreilly.com/ideas/why-2016-is-shaping-up-to-be-the-year-of-the-bot.

39 Kareem Anderson, "Microsoft CEO Satya Nadella Says Chatbots Will Revolutionize Computing," *On MSFT*, July 11, 2016, https://www.onmsft.com/news/microsoft-ceo-satya-nadella-says-chatbots-will-revolutionize-computing.

40 Ann Ravel, Hamsini Sridharan, and Samuel Woolley, "Principles and Policies to Counter Deceptive Digital Politics," Maplight and the Institute for the Future, February 12, 2019, https://s3-us-west-2.amazonaws.com/maplight.org/wp-content/uploads/20190211224524/Principles-and-Policies-to-Counter-Deceptive-Digital-Politics-1-1-2.pdf.

41 Nadine Strossen, *HATE: Why We Should Resist It with Free Speech, Not Censorship* (New York: Oxford University Press, 2018).

42 Jessica Leinwand, "Expanding Our Policies on Voter Suppression," *Facebook Newsroom*, October 15, 2018, https://newsroom.fb.com/news/2018/10/voter-suppression-policies/ (accessed February 13, 2019).

43 Nancy Scola, "Experts Warn the Social Media Threat This Election Is Homegrown," *Politico*, November 5, 2018, https://politi.co/2ANTw0T.

44 Gabriel J. X. Dance, Michael LaForgia, and Nicholas Confessore, "As Facebook Raised a Privacy Wall, It Carved an Opening for Tech Giants," *New York Times*, December 18, 2018, https://www.nytimes.com/2018/12/18/technology/facebook-privacy.html.

45 Andrew Arnold, "Do We Really Need to Start Regulating Social Media?," *Forbes*, July 30, 2018, https://www.forbes.com/sites/andrewarnold/2018/07/30/do-we-really-need-to-start-regulating-social-media/; Amelia Irvine, "Don't Regulate Social Media Companies—Even if They Let Holocaust Deniers Speak," *USA Today*, July 19, 2018, https://www.usatoday.com/story/opinion/2018/07/19/dont-regulate-social-media-despite-bias-facebook-twitter-youtube-column/796471002/.

Chapter Four

1 Kurt Wagner, "Mark Zuckerberg Says It's 'Crazy' to Think Fake News Stories Got Donald Trump Elected," *Vox*, November 11, 2016, https://www.vox.com/2016/11/11/13596792/facebook-fake-news-mark-zuckerberg-donald-trump.

2 Chris Prentice, "Zuckerberg Again Rejects Claims of Facebook Impact on US Election," *Reuters*, November 13, 2016, https://www.reuters.com/article/us-usa-election-facebook-idUSKBN1380TH.

3 Sheera Frenkel et al., "Delay, Deny, and Deflect: How Facebook's Leaders Fought through Crisis," *New York Times*, November 30, 2018, https://www.nytimes.com/2018/11/14/technology/facebook-data-russia-election-racism.html.

4 Carla Herreria, "Mark Zuckerberg: 'I Regret' Rejecting Idea That Facebook Fake News Altered Election," *Huffington Post*, September 28, 2017, https://www.huffingtonpost.com/entry/mark-zuckerberg-regrets-fake-news-facebook_us_59cc2039e4b05063fe0eed9d.

5 Shannon Liao, "11 Weird and Awkward Moments from Two Days of Mark Zuckerberg's Congressional Hearing," *The Verge*, April 11, 2018, https://www.theverge.com/2018/4/11/17224184/facebook-mark-zuckerberg-congress-senators.

6 Drew Harwell, "AI Will Solve Facebook's Most Vexing Problems, Mark Zuckerberg Says. Just Don't Ask When or How," *Washington Post*, April 11, 2018, https://www.washingtonpost.com/news/the-switch/wp/2018/04/11/ai-will-solve-facebooks-most-vexing-problems-mark-zuckerberg-says-just-dont-ask-when-or-how/.

7 Tom Simonite, "AI Has Started Cleaning Up Facebook, but Can It Finish?," *Wired*, December 18, 2018, https://www.wired.com/story/ai-has-started-cleaning-facebook-can-it-finish/.

8 Harwell, "AI Will Solve Facebook's Most Vexing Problems, Mark Zuckerberg Says."

9 Will Knight, "A New AI Algorithm Summarizes Text Amazingly Well," *MIT Technology Review*, May 12, 2017, https://www.technologyreview.com/s/607828/an-algorithm-summarizes-lengthy-text-surprisingly-well/; Kyle Wiggers, "Microsoft Develops Flexible AI System That Can Summarize the News," *VentureBeat*, November 6, 2018, https://venturebeat.com/2018/11/06/microsoft-researchers-develop-ai-system-that-can-generate-articles-summaries/.

10 Katyanna Quach, "Look Out, Wiki-Geeks. Now Google Trains AI to Write Wikipedia Articles," *The Register*, February 15, 2018, https://www.theregister.co.uk/2018/02/15/google_brain_ai_wikipedia/.

11 Tanza Loudenback, Melissa Stranger, and Emmie Martin, "The 13 Richest People in Tech," *Business Insider*, February 3, 2016, https://www.businessinsider.com/richest-people-in-tech-2016-1.

12 "GDP (current US$)," World Bank national accounts data, https://data.worldbank.org/indicator/NY.GDP.MKTP.CD?view=map (accessed February 13, 2019); "Asian and European Cities Compete for the Title of Most Expensive City," *The Economist*, March 15, 2018, https://www.economist.com/graphic-detail/2018/03/15/asian-and-european-cities-compete-for-the-title-of-most-expensive-city.

13 Scott Shackelford, "Facebook's Social Responsibility Should Include Privacy Protection," *The Conversation*, April 12, 2018, http://theconversation.com/facebooks-social-responsibility-should-include-privacy-protection-94549.

14 Berit Anderson and Brett Horvath, "The Rise of the Weaponized AI Propaganda Machine," *Scout*, February 9, 2017, https://scout.ai/story/the-rise-of-the-weaponized-ai-propaganda-machine.

15 Vyacheslav Polonski, "How Artificial Intelligence Conquered Democracy," *The Conversation*, August 8, 2017, http://theconversation.com/how-artificial-intelligence-conquered-democracy-77675.

16 Vyacheslav Polonski, "Artificial Intelligence Has the Power to Destroy or Save Democracy," Council on Foreign Relations, *Net Politics*, August 7, 2017, https://www.cfr.org/blog/artificial-intelligence-has-power-destroy-or-save-democracy.

17 Samuel Woolley, "Manufacturing Consensus: Computational Propaganda and the 2016 United States Presidential Election," PhD diss., University of Washington (2018).

18 Philip N. Howard and Bence Kollanyi, "Bots, #StrongerIn, and #Brexit: Computational Propaganda during the UK-EU Referendum," Cornell University, Computer Science: Social and Information Networks, June 20, 2016, http://arxiv.org/abs/1606.06356.

19 Shawn Musgrave, "'I Get Called a Russian Bot 50 Times a Day,'" *Politico Magazine*, August 9, 2017, http://politi.co/2viAxZA.

20 Chris Elliott, "The Readers' Editor on...Pro-Russia Trolling below the Line on Ukraine Stories," *Guardian*, May 4, 2014, https://www.theguardian.com/commentisfree/2014/may/04/pro-russia-trolls-ukraine-guardian-online.

21 Juliana Gragnani, "Inside the World of Brazil's Social Media Cyborgs," December 13, 2017, https://www.bbc.com/news/world-latin-america-42322064.

22 Samuel Woolley and Katie Joseff, "Computational Propaganda and the 2018 US Midterms: Executive Summary," Institute for the Future, Digital Intelligence Lab working paper (forthcoming).

23 Jeremy Hsu, "Why 'Uncanny Valley' Human Look-Alikes Put Us on Edge," *Scientific American*, March 4, 2012, https://www.scientificamerican.com/article/why-uncanny-valley-human-look-alikes-put-us-on-edge/.

24 Kevin Kelly, *The Inevitable: Understanding the 12 Technological Forces That Will Shape Our Future* (New York: Penguin Books, 2017), 13.

25 Lisa-Maria Neudert, "Future Elections May Be Swayed by Intelligent, Weaponized Chatbots," *MIT Technology Review*, August 22, 2018, https://www.technologyreview.com/s/611832/future-elections-may-be-swayed-by-intelligent-weaponized-chatbots/.

26 George Dvorsky, "Hackers Have Already Started to Weaponize Artificial Intelligence," *Gizmodo*, https://gizmodo.com/hackers-have-already-started-to-weaponize-artificial-in-1797688425 (accessed May 1, 2019).

27 Matthew Jagielski, Alina Oprea, Battista Biggio, Chang Liu, Cristina Nita-Rotaru, and Bo Li, "Manipulating Machine Learning: Poisoning Attacks and Countermeasures for Regression Learning," Cornell University, Computer Science: Cryptography and Security, April 1, 2018, http://arxiv.org/abs/1804.00308.

28 Kalev Leetaru, "What if Deep Learning Was Given Command of a Botnet?," *Forbes*, January 11, 2017, https://www.forbes.com/sites/kalevleetaru/2017/01/11/what-if-deep-learning-was-given-command-of-a-botnet/.

29 Gaby Galvin, "How Bots Could Hack Your Health," *US News & World Report*, July 24, 2018, https://www.usnews.com/news/healthiest-communities/articles/2018-07-24/how-social-media-bots-could-compromise-public-health.

30 Marc Owen Jones, "Someone in Saudi Has Taken over the Verified Twitter Account of a Dead American Meteorologist—and There's More...," February 9, 2019, https://marcowenjones.wordpress.com/2019/02/09/someone-in-saudi-has-taken-over-the-verified-twitter-account-of-a-dead-american-meteorologist-and-theres-more/.

31 OSoMe, "Botometer: An OSoMe Project," https://botometer.iuni.iu.edu.

32 Tessa Lyons, "Increasing Our Efforts to Fight False News," *Facebook Newsroom*, June 21, 2018, https://newsroom.fb.com/news/2018/06/increasing-our-efforts-to-fight-false-news/.

33 Petter Bae Brandtzaeg, Asbjørn Følstad, and María Ángeles Chaparro Domínguez, "How Journalists and Social Media Users Perceive Online Fact-Checking and Verification Services," *Journalism Practice* 12, no. 9 (October 21, 2018): 1109–1129, https://doi.org/10.1080/17512786.2017.1363657; R. Kelly Garrett, Erik C. Nisbet, and Emily K. Lynch, "Undermining the Corrective Effects of Media-Based Political Fact Checking? The Role of Contextual Cues and Naive Theory," *Journal of Communication* 63, no. 4 (August 1, 2013): 617–637, https://doi.org/10.1111/jcom.12038.

34 Daniel Funke, "Snopes Pulls Out of Its Fact-Checking Partnership with Facebook," *Poynter*, February 1, 2019, https://www.poynter.org/fact-checking/2019/snopes-pulls-out-of-its-fact-checking-partnership-with-facebook/.

35 Sandra E. Garcia, "Ex-Content Moderator Sues Facebook, Saying Violent Images Caused Her PTSD," *New York Times*, December 28, 2018, https://www.nytimes.com/2018/09/25/technology/facebook-moderator-job-ptsd-lawsuit.html.

36 Casey Newton, "The Trauma Floor: The Secret Lives of Facebook Moderators in America," *The Verge*, February 25, 2019, https://www.theverge.com/2019/2/25/18229714/cognizant-facebook-content-moderator-interviews-trauma-working-conditions-arizona.

37 Nicole Karlis, "Facebook's Shift to 'Privacy-Focused' Company: Earnest Change or Cynical PR Move?," *Salon*, March 8, 2019, www.salon.com/2019/03/07/facebooks-shift-to-privacy-company-earnest-change-or-cynical-pr-move/.

38 James Vincent, "Facebook Is Using Machine Learning to Spot Hoax Articles Shared by Spammers," *The Verge*, June 21, 2018, https://www.theverge.com/2018/6/21/17488040/facebook-machine-learning-spot-hoax-articles-spammers.

39 Eliza Strickland, "AI-Human Partnerships Tackle 'Fake News,'" *IEEE Spectrum: Technology, Engineering, and Science News*, August 29, 2018, https://spectrum.ieee.org/computing/software/aihuman-partnerships-tackle-fake-news.

40 Terry Collins, "Facebook Vows to Do Better Combating Hate Speech," *CNET*, June 27, 2017, https://www.cnet.com/news/facebook-its-hard-handling-hate-speech/.

41 Emily Sullivan, "Twitter Says It Will Ban Threatening Accounts, Starting Today," *The Two-Way*, NPR, December 18, 2017, https://www.npr.org/sections/thetwo-way/2017/12/18/571622652/twitter-says-it-will-ban-threatening-accounts-starting-today.

42 Brett Samuels, "Facebook Apologizes to Texas Newspaper for Part of Declaration of Independence Being Labeled Hate Speech," *The Hill*, July 5, 2018, https://thehill.com/policy/technology/395583-facebook-apologizes-for-labeling-part-of-declaration-of-independence-as.

43 Sam Wolfson, "Facebook Labels Declaration of Independence as 'Hate Speech,'" *Guardian*, July 5, 2018, https://www.theguardian.com/world/2018/jul/05/facebook-declaration-of-independence-hate-speech.

44 Ali Breland, "How White Engineers Built Racist Code—and Why It's Dangerous for Black People," the *Guardian*, December 4, 2017, sec. Technology, https://www.theguardian.com/technology/2017/dec/04/racist-facial-recognition-white-coders-black-people-police.

45 Christian Sandvig, Kevin Hamilton, Karrie Karahalios, and Cedric Langbort, "Automation, Algorithms, and Politics: When the Algorithm Itself Is a Racist: Diagnosing Ethical Harm in the Basic Components of Software," *International Journal of Communication* 10 (October 12, 2016): 19; James Zou and Londa Schiebinger, "AI Can Be Sexist and Racist—It's Time to Make It Fair," *Nature* 559, no. 7714 (July 2018): 324, https://doi.org/10.1038/d41586-018-05707-8.

46 Oren Etzioni, "No, the Experts Don't Think Superintelligent AI Is a Threat to Humanity," *MIT Technology Review*, September 20, 2016, https://www.technologyreview.com/s/602410/no-the-experts-dont-think-superintelligent-ai-is-a-threat-to-humanity/.

47 Andrew Myers, "Artificial Intelligence Index Tracks Emerging Field," *Stanford News*, November 30, 2017, https://news.stanford.edu/2017/11/30/artificial-intelligence-index-tracks-emerging-field/.

48 Christopher Mims, "Without Humans, Artificial Intelligence Is Still Pretty Stupid," *Wall Street Journal*, November 12, 2017, https://www.wsj.com/articles/without-humans-artificial-intelligence-is-still-pretty-stupid-1510488000.

49 Grady Booch, "Don't Fear Superintelligent AI," TED Talk, November 2016, https://www.ted.com/talks/grady_booch_don_t_fear_superintelligence.

Chapter Five

1 Kyle Swenson, "How CNN's Jim Acosta Became the Reporter Trump Loves to Hate," *Washington Post*, November 8, 2018, https://www.washingtonpost.com/nation/2018/11/08/how-cnns-jim-acosta-became-reporter-trump-loves-hate/.

2 Erin Durkin and Ben Jacobs, "White House Backs Down in Fight with CNN over Jim Acosta—as It Happened," *Guardian*, November 19, 2018, https://www.theguardian.com/us-news/live/2018/nov/19/white-house-jim-acosta-trump-latest-us-politics-live.

3 Naomi Lim, "CNN's Acosta Denies 'Placing His Hands on' White House Intern," *Washington Examiner*, November 8, 2018, https://www.washingtonexaminer.com/news/cnns-acosta-denies-placing-his-hands-on-white-house-intern.

4 Associated Press, "White House Bans CNN Reporter Jim Acosta after a Confrontation with Trump," CNBC, November 8, 2018, https://www.cnbc.com/2018/11/08/white-house-bans-cnn-reporter-jim-acosta-after-a-confrontation-with-trump-.html.

5 Drew Harwell, "White House Shares Doctored Video to Support Punishment of Journalist Jim Acosta," *Washington Post*, October 8, 2018, https://www.washingtonpost.com/technology/2018/11/08/white-house-shares-doctored-video-support-punishment-journalist-jim-acosta/.

6 Allison Chiu, "Kellyanne Conway on Jim Acosta Video: 'That's Not Altered. That's Sped Up. They Do It All the Time in Sports,'" *Washington Post*, November 12, 2018, https://www.washingtonpost.com/nation/2018/11/12/kellyanne-conway-acosta-video-thats-not-altered-thats-sped-up-they-do-it-all-time-sports/.

7 "The Big Question: How Will 'Deepfakes' and Emerging Technology Transform Disinformation?" National Endowment for Democracy, October 1, 2018, https://www.ned.org/the-big-question-how-will-deepfakes-and-emerging-technology-transform-disinformation/.

8 Drew Harwell, "Fake-Porn Videos Are Being Weaponized to Harass and Humiliate Women: 'Everybody Is a Potential Target,'" *Washington Post*, December 30, 2018, https://www.washingtonpost.com/technology/2018/12/30/fake-porn-videos-are-being-weaponized-harass-humiliate-women-everybody-is-potential-target/.

9 Cade Metz, "Google Just Open Sourced the Artificial Intelligence Engine at the Heart of Its Online Empire," *Wired*, November 9, 2015, https://www.wired.com/2015/11/google-open-sources-its-artificial-intelligence-engine/.

10 Tom Simonite, "Will 'Deepfakes' Disrupt the Midterm Election?," *Wired*, November 1, 2018, https://www.wired.com/story/will-deepfakes-disrupt-the-midterm-election/.

11 Oscar Schwartz, "You Thought Fake News Was Bad? Deep Fakes Are Where Truth Goes to Die," *Guardian*, November 12, 2018, https://www.theguardian.com/technology/2018/nov/12/deep-fakes-fake-news-truth.

12 Samantha Cole and Emanuel Maiberg, "Pornhub Is Banning AI-Generated Fake Porn Videos, Says They're Nonconsensual," *Motherboard*, February 6, 2018, https://motherboard.vice.com/en_us/article/zmwvdw/pornhub-bans-deepfakes.

13 James Vincent, "Watch Jordan Peele Use AI to Make Barack Obama Deliver a PSA about Fake News," *The Verge*, April 17, 2018, https://www.theverge.com/tldr/2018/4/17/17247334/ai-fake-news-video-barack-obama-jordan-peele-buzzfeed.

14 Kim Hyeongwoo et al., "Deep Video Portraits," in *ACM* [Association for Computing Machinery]*: Transactions of Graphics* 37, no. 4 (August 2018).

15 Jennifer Langston, "Lip-Syncing Obama: New Tools Turn Audio Clips into Realistic Video," *UW News*, July 11, 2017, http://www.washington.edu/news/2017/07/11/lip-syncing-obama-new-tools-turn-audio-clips-into-realistic-video/.

16 "A Faked Video of Donald Trump Points to a Worrying Future," *The Economist*, May 24, 2018, https://www.economist.com/leaders/2018/05/24/a-faked-video-of-donald-trump-points-to-a-worrying-future.

17 Tim Hwang, "Don't Worry about Deepfakes. Worry about Why People Fall for Them," *Undark*, December 13, 2018, https://undark.org/2018/12/13/how-worried-should-we-be-about-deepfakes/.

18 Chris Cillizza, "Why Fact-Checking Doesn't Change People's Minds," *Washington Post*, February 23, 2017, https://www.washingtonpost.com/news/the-fix/wp/2017/02/23/why-fact-checking-doesnt-change-peoples-minds/.

19 Brendan Nyhan and Jason Reifler, "When Corrections Fail: The Persistence of Political Misperceptions," *Political Behavior* 32, no. 2 (June 1, 2010): 303–330, https://doi.org/10.1007/s11109-010-9112-2.

20 Kevin Roose, "Here Come the Fake Videos, Too," *New York Times*, June 8, 2018, https://www.nytimes.com/2018/03/04/technology/fake-videos-deepfakes.html.

21 Tom Gara, "It's Not Fake Video We Should Be Worried About, It's Real Video," *BuzzFeed*, January 24, 2019, https://www.buzzfeednews.com/article/tomgara/fake-video-isnt-the-future-of-propaganda-real-video-works.

22 Limelight Networks, "The State of Online Video 2018," https://www.limelight.com/resources/white-paper/state-of-online-video-2018.

23 Cisco, "Cisco Visual Networking Index: Forecast and Trends, 2017–2022: White Paper," updated February 27, 2019, https://www.cisco.com/c/en/us/solutions/collateral/service-provider/visual-networking-index-vni/white-paper-c11-741490,html (accessed February 13, 2019).

24 Gara, "It's Not Fake Video We Should Be Worried About."

25 Robert Mackey, "Fake Interview with Alexandria Ocasio-Cortez Was Satire, Not Hoax, Conservative Pundit Says," *Intercept*, July 24, 2018, https://theintercept.com/2018/07/24/conservative-network-says-fake-interview-alexandria-ocasio-cortez-satire/.

26 Alexandria Ocasio-Cortez (*@AOC*), tweet of July 24, 2018, https://twitter.com/AOC/status/1021750530249568257?ref_src=twsrc%5Etfw%7Ctwcamp%5Etweetembed%7Ctwterm%5E1021750530249568257&ref_url=https%3A%2F%2Ftheintercept.com%2F2018%2F07%2F24%2Fconservative-network-says-fake-interview-alexandria-ocasio-cortez-satire%2F.

27 3M, "Polishing Your Presentation," 3M Meeting Network, November 2, 2000, http://web.archive.org/web/20001102203936/http%3A//3m.com/meetingnetwork/files/meetingguide_pres.pdf.

28 Christina J. Howard and Alex O. Holcombe, "Unexpected Changes in Direction of Motion Capture Attention," *Attention, Perception, and Psychophysics* 72, no. 8 (2010): 2087–2095, https://doi.org/doi:10.3758/APP.72.8.2087.

29 Melanie Green and Timothy Brock, "The Role of Transportation in the Persuasiveness of Public Narratives," *Journal of Personality and Social Psychology* 79, no. 5 (2000): 701.

30 Kelly Born, "Disinformation Threatens 2020 Election," *Morning Consult*, April 29, 2019, https://morningconsult.com/opinions/disinformation-threatens-2020-election/ (accessed May 1, 2019).

31 Rebecca Lewis, "Alternative Influence: Broadcasting the Reactionary Right on YouTube," Data & Society Research Institute, September 18, 2018, https://datasociety.net/wp-content/uploads/2018/09/DS_Alternative_Influence.pdf.

32 Philip Schindler, "The Google News Initiative: Building a stronger future for news", March 20, 2018, https://blog.google/outreach-initiatives/google-news-initiative/announcing-google-news-initiative/.

33 Michael Nuñez, "YouTube Announces Sweeping Changes to the Way It Handles Breaking News," *Mashable*, August 9, 2018, https://mashable.com/article/youtube-announces-changes-breaking-news-video-search/.

34 Kreiss and McGregor, "Technology Firms Shape Political Communication."

35 Louise Matsakis, "YouTube Will Link Directly to Wikipedia to Fight Conspiracy Theories," *Wired*, March 13, 2018, https://www.wired.com/story/youtube-will-link-directly-to-wikipedia-to-fight-conspiracies/.

36 John Herrman, "YouTube May Add to the Burdens of Humble Wikipedia," *New York Times*, June 8, 2018, https://www.nytimes.com/2018/03/19/business/media/youtube-wikipedia.html.

37 Vinny Green and David Mikkelson, "A Message to Our Community Regarding the Facebook Fact-Checking Partnership," Snopes, February 1, 2019, https://www.snopes.com/snopes-fb-partnership-ends/.

38 Sarah T. Roberts, "Commercial Content Moderation: Digital Laborers' Dirty Work," in *The Intersectional Internet: Race, Sex, Class, and Culture Online*, ed. Safiya Umoja Noble and Brendesha M. Tyne (New York: Peter Lang Publishing, 2016).

39 Yuezun Li, Ming-Ching Chang, and Siwei Lyu, "In Ictu Oculi: Exposing AI Generated Fake Face Videos by Detecting Eye Blinking," Cornell University, Computer Science: Computer Vision and Pattern Recognition, June 7, 2018, http://arxiv.org/abs/1806.02877.

40 Siwei Lyu, "Detecting 'Deepfake' Videos in the Blink of an Eye," *The Conversation*, August 29, 2018, http://theconversation.com/detecting-deepfake-videos-in-the-blink-of-an-eye-101072.

41 Francesco Marconi and Till Daldrup, "How the Wall Street Journal Is Preparing Its Journalists to Detect Deepfakes," *Nieman Lab*, November 15, 2018, http://www.niemanlab.org/2018/11/how-the-wall-street-journal-is-preparing-its-journalists-to-detect-deepfakes/.

42 Lily Hay Newman, "A New Tool Protects Videos from Deepfakes and Tampering," *Wired*, February 11, 2019, https://www.wired.com/story/amber-authenticate-video-validation-blockchain-tampering-deepfakes/.

43 Matthew Field, "Hacker Claims He Will Live-Stream Deleting Zuckerberg's Facebook Profile," *Telegraph*, September 28, 2018, https://www.telegraph.co.uk/technology/2018/09/28/hacker-claims-will-live-stream-deleting-zuckerbergs-facebook/.

44 Helen Chen, "Fortnite Gamer Accused of Live Streaming Domestic Violence Assault Granted Bail," *SBS Your Language*, November 12, 2018, https://www.sbs.com.au/yourlanguage/mandarin/en/article/2018/12/11/fortnite-gamer-accused-live-streaming-domestic-violence-assault-granted-bail.

45 Kathleen Chaykowski, "Terrorism Suspect's Use of 'Facebook Live' Stream Highlights Challenges of Live Video," *Forbes*, June 15, 2016, https://www.forbes.com/sites/kathleenchaykowski/2016/06/15/terrorist-suspects-use-of-facebook-live-stream-highlights-challenges-of-live-video/.

46 Bruce Sterling, "Disinformation Digest," *Wired*, November 20, 2016, https://www.wired.com/beyond-the-beyond/2016/11/disinformation-digest/.

47 Sheryl Sandberg, "Facebook Chief Operating Officer Sheryl Sandberg's Letter to New Zealand," *New Zealand Herald*, March 30, 2019, https://www.nzherald.co.nz/business/news/article.cfm?c_id=3&objectid=12217454.

48 Mark Zuckerberg has said publicly that the future of the company is private and that Facebook will move away from the newsfeed model. Some fear that the move to privacy will simply create more walled gardens that researchers and law enforcement cannot monitor. See Nick Statt, "Facebook CEO Mark Zuckerberg

Says the 'Future Is Private,'" *The Verge*, April 30, 2019, https://www.theverge.com/2019/4/30/18524188/facebook-f8-keynote-mark-zuckerberg-privacy-future-2019.

Chapter Six

1 Jakob Verbruggen, "Men Against Fire," *Black Mirror*, Netflix, October 10, 2016, http://www.imdb.com/title/tt5709234/.
2 "Kengoro, the Most Advanced Humanoid Robot Yet," *RobotsVoice*, March 8, 2018, http://www.robotsvoice.com/kengoro-advanced-humanoid-robot-yet/.
3 For more on IFTF and the Emerging Media Lab, visit iftf.org.
4 Kangpan, "Bright Spots in the VR Market."
5 Kevin Kelly, "AR Will Spark the Next Big Tech Platform—Call It Mirrorworld," *Wired*, February 12, 2019, https://www.wired.com/story/mirrorworld-ar-next-big-tech-platform/.
6 David Gelernter, *Mirror Worlds: Or the Day Software Puts the Universe in a Shoebox...How It Will Happen and What It Will Mean* (New York: Oxford University Press, 1993).
7 Emory Craig, "Marc Andreessen: VR Will Be '1,000' Times Bigger than AR," *Digital Bodies*, January 7, 2019, https://www.digitalbodies.net/virtual-reality/marc-andreessen-vr-will-be-1000-times-bigger-than-ar/.
8 Rebecca Hills-Duty, "China's Communist Party Uses VR for Loyalty Tests," *VRFocus*, May 9, 2018, https://www.vrfocus.com/2018/05/chinas-communist-party-uses-vr-for-loyalty-tests/.
9 Gideon Resnick and Ben Collins, "Palmer Luckey: The Facebook Near-Billionaire Secretly Funding Trump's Meme Machine," *Daily Beast*, September 23, 2016, https://www.thedailybeast.com/articles/2016/09/22/palmer-luckey-the-facebook-billionaire-secretly-funding-trump-s-meme-machine.
10 Nick Wingfield, "Oculus Founder Plots a Comeback with a Virtual Border Wall," *New York Times*, December 22, 2017, https://www.nytimes.com/2017/06/04/business/oculus-palmer-luckey-new-start-up.html.
11 Tom Huddleston Jr., "Oculus Co-Founder Palmer Luckey Wants to Build a 'Virtual' Border Wall," *CNBC Make It*, January 15, 2019, https://www.cnbc.com/2019/01/15/oculus-co-founder-palmer-luckey-wants-to-build-a-virtual-border-wall.html.

12 "Hate in Social VR," Anti-Defamation League, https://www.adl.org/resources/reports/hate-in-social-virtual-reality (accessed February 18, 2019).

13 Jessica Outlaw, "Virtual Harassment: The Social Experience of 600+ Regular Virtual Reality (VR) Users," *The Extended Mind*, April 4, 2018, https://extendedmind.io/blog/2018/4/4/virtual-harassment-the-social-experience-of-600-regular-virtual-reality-vrusers.

14 Casey Newton, "People Older than 65 Share the Most Fake News, a New Study Finds," *The Verge*, January 9, 2019, https://www.theverge.com/2019/1/9/18174631/old-people-fake-news-facebook-share-nyu-princeton.

15 Samuel Woolley and Philip Howard, "Computational Propaganda: Executive Summary," University of Oxford, Oxford Internet Institute working paper (June 2017).

16 Anti-Defamation League, "Reverend Patricia Novick, PhD, DMin," https://www.adl.org/reverend_patricia_novick%2C_ph.d.%2C_d.min (accessed February 19, 2019).

17 Andy Brownstone, "Can Virtual Reality Reduce Racism?" *BBC News*, November 28, 2013, https://www.bbc.com/news/av/science-environment-24929089/reducing-ingrained-racism-with-virtual-reality.

18 Tyler Young and Ankita Rao, "This VR Founder Wants to Gamify Empathy to Reduce Racial Bias," *Motherboard*, July 20, 2018, https://motherboard.vice.com/en_us/article/a3qeyk/this-vr-founder-wants-to-gamify-empathy-to-reduce-racial-bias.

19 Zillah Watson, "VR for News: The New Reality?," Reuters Institute for the Study of Journalism, Digital News Project, 2017, https://reutersinstitute.politics.ox.ac.uk/our-research/vr-news-new-reality.

20 Umberto Bacchi, "On the Frontline of Climate Change in the South Pacific," *Independent*, July 7, 2017, http://www.independent.co.uk/news/world/australasia/climate-change-south-pacific-global-warming-sea-levels-a7829786.html.

21 Jeremy Bailenson, "Virtual Reality Can Help Politicians Make Responsible Decisions about the Environment," *National Geographic Society Newsroom*, October 25, 2017, https://blog.nationalgeographic.org/2017/10/25/virtual-reality-can-help-politicians-make-responsible-decisions-about-the-environment/.

22 Bailenson, "Virtual Reality Can Help Politicians Make Responsible Decisions."

23 David Remnick, "Episode 87: Virtual Reality, and the Politics of Genetics," *New Yorker*, June 16, 2017, https://www.newyorker.com/podcast/the-new-yorker-radio-hour/episode-87-virtual-reality-and-the-politics-of-genetics.
24 Janet Murray, "Who's Afraid of the Holodeck? Facing the Future of Digital Narrative without Ludoparanoia," lecture delivered May 22, 2017, University of Utrecht and HKU Interactive Narrative Design (the Netherlands), https://www.youtube.com/watch?v=zQpaM0kEf70.
25 Several of these principles are drawn from Ravel, Sridharan, and Woolley, "Principles and Policies to Counter Deceptive Digital Politics."
26 Michael Santoli, "It Could Become 'Facebook Thursday,' Akin to Infamous 'Marlboro Friday' Plunge," CNBC, July 26, 2018, https://www.cnbc.com/2018/07/26/facebook-shares-may-rebound-after-plunge-just-likephilip-morris-in-ea.html.

Chapter Seven

1 Irving John Good, "Speculations Concerning the First Ultraintelligent Machine," *Advances in Computers* 6 (1966): 31–88, https://doi.org/10.1016/S0065-2458(08)60418-0.
2 Marlow Stern and Jen Yamato, "SXSW 2016's Biggest Stars: President Obama, Atlanta Hip-Hop, and More," *Daily Beast*, March 20, 2016, https://www.thedailybeast.com/articles/2016/03/20/sxsw-2016-s-biggest-stars-president-obama-atlanta-hip-hop-and-more.
3 "RoboPresident: Politics in an Algorithmic World," SXSW Schedule 2016, https://schedule.sxsw.com/2016/events/event_PP49943 (accessed February 23, 2019).
4 Tad Friend, "How Frightened Should We Be of AI?," *New Yorker*, May 7, 2018, https://www.newyorker.com/magazine/2018/05/14/how-frightened-should-we-be-of-ai.
5 Monica Nickelsburg, "That's One Smooth-Talking Robot: Google's WaveNet AI Program Produces Human-like Speech," *GeekWire*, September 12, 2016, https://www.geekwire.com/2016/thats-one-smooth-talking-robot-googles-wavenet-ai-program-produces-human-like-speech/.
6 Jessi Hempel, "Siri and Cortana Sound Like Ladies Because of Sexism," *Wired*, October 28, 2015, https://www.wired.com/2015/10/why-siri-cortana-voice-interfaces-sound-female-sexism/.

7 Danielle De La Bastide, "Researchers at MIT Just Created a Very Polite Robot," *Interesting Engineering*, September 4, 2017, https://interestingengineering.com/researchers-at-mit-just-created-a-very-polite-robot.
8 Aviva Rutkin, "Not Like Us: How Should We Treat the Robots We Live Alongside?," *New Scientist*, October 6, 2015, https://www.newscientist.com/article/dn28293-not-like-us-how-should-we-treat-the-robots-we-live-alongside/.
9 Larry Bartleet, "Swearing at Your Phone Gets You Better Customer Service in America," *NME*, November 22, 2017, https://www.nme.com/blogs/nme-blogs/hilarious-hack-americans-never-waste-time-automated-phone-menus-2162409.
10 Booch, "Don't Fear Superintelligent AI."
11 See in particular Juliana Schroeder and Nicholas Epley, "The Sound of Intellect: Speech Reveals a Thoughtful Mind, Increasing a Job Candidate's Appeal," *Psychological Science* 26, no. 6 (June 1, 2015): 877–891, https://doi.org/10.1177/0956797615572906; Juliana Schroeder and Nicholas Epley, "Mistaking Minds and Machines: How Speech Affects Dehumanization and Anthropomorphism," *Journal of Experimental Psychology: General* 145, no. 11 (2016): 1427–1437.
12 David Pierce, "How Apple Finally Made Siri Sound More Human," *Wired*, September 7, 2017, https://www.wired.com/story/how-apple-finally-made-siri-sound-more-human/.
13 Olivia Solon, "Google's Robot Assistant Now Makes Eerily Lifelike Phone Calls for You," *Guardian*, May 8, 2018, https://www.theguardian.com/technology/2018/may/08/google-duplex-assistant-phone-calls-robot-human.
14 Yongdong Wang, "The Chatbot That's Acing the Largest Turing Test in History," *Nautilus*, February 4, 2016, http://nautil.us/issue/33/attraction/your-next-new-best-friend-might-be-a-robot (accessed August 12, 2016).
15 Solon, "Google's Robot Assistant Now Makes Eerily Lifelike Phone Calls for You."
16 "You'll Want to Keep an Eye on These 10 Breakthrough Technologies This Year," *MIT Technology Review* (March/April 2018), https://www.technologyreview.com/lists/technologies/2018/.
17 Ian J. Goodfellow, Jean Pouget-Abadie, Mehdi Mirza, Bing Xu, David Warde-Farley, Sherjil Ozair, Aaron Courville, and Yoshua Bengio, "Generative Adversarial Networks," Cornell University, Statistics: Machine Learning, June 10, 2014, http://arxiv.org/abs/1406.2661; Vincent Dumoulin, Jonathon

Shlens, and Manjunath Kudlur, "Supercharging Style Transfer," *Google AI Blog*, December 26, 2016, http://ai.googleblog.com/2016/10/supercharging-style-transfer.html.

18 Jordan Pearson and Natasha Grzincic, "These People Are Not Real—They Were Created by AI," *Motherboard*, December 14, 2018, https://motherboard.vice.com/en_us/article/mby4q8/these-people-were-created-by-nvidia-ai.

19 James Vincent, "These Faces Show How Far AI Image Generation Has Advanced in Just Four Years," *The Verge*, December 17, 2018, https://www.theverge.com/2018/12/17/18144356/ai-image-generation-fake-faces-people-nvidia-generative-adversarial-networks-gans.

20 Jack Nicas, "Facebook Says Russian Firms 'Scraped' Data, Some for Facial Recognition," *New York Times*, November 26, 2018, https://www.nytimes.com/2018/10/12/technology/facebook-russian-scraping-data.html.

21 Natasha Singer, "Amazon's Facial Recognition Wrongly Identifies 28 Lawmakers, ACLU Says," *New York Times*, July 27, 2018, https://www.nytimes.com/2018/07/26/technology/amazon-aclu-facial-recognition-congress.html.

22 Jacob Snow, "Amazon's Face Recognition Falsely Matched 28 Members of Congress with Mugshots," American Civil Liberties Union, July 26, 2018, https://www.aclu.org/blog/privacy-technology/surveillance-technologies/amazons-face-recognition-falsely-matched-28.

23 Joy Buolamwini and Timnit Gebru, "Gender Shades: Intersectional Accuracy Disparities in Commercial Gender Classification," *Proceedings of the First Conference on Fairness, Accountability, and Transparency, Proceedings of Machine Learning Research* 81 (2018): 77–91, http://proceedings.mlr.press/v81/buolamwini18a.html.

Chapter Eight

1 "Data Never Sleeps 5.0," DOMO, 2018, https://www.domo.com/learn/data-never-sleeps-5.

2 Bernard Marr, "How Much Data Do We Create Every Day? The Mind-Blowing Stats Everyone Should Read," *Forbes*, May 21, 2018, https://www.forbes.com/sites/bernardmarr/2018/05/21/how-much-data-do-we-create-every-day-the-mind-blowing-stats-everyone-should-read/.

3 "Hamilton 68 Version 2.0," Alliance for Securing Democracy, https://securingdemocracy.gmfus.org/hamilton-68/.

4 Samuel Woolley and Marina Gorbis, "Social Media Bots Threaten Democracy. But We Are Not Helpless," *Guardian*, October 16, 2017, https://www.theguardian.com/commentisfree/2017/oct/16/bots-social-media-threaten-democracy-technology.

5 danah boyd, "Did Media Literacy Backfire?," *Data & Society: Points*, January 5, 2017, https://points.datasociety.net/did-media-literacy-backfire-7418c084d88d.

6 Social Science One, https://socialscience.one/home (accessed February 11, 2019).

7 Fruzsina Eordogh, "YouTube Stops Recommending Conspiracy Videos, Finally," *Forbes*, January 28, 2019, https://www.forbes.com/sites/fruzsinaeordogh/2019/01/28/youtube-stops-recommending-conspiracy-videos-finally/.

8 Kreiss and McGregor, "Technology Firms Shape Political Communication."

9 Sam Greenspan, Bellwether, https://bellwether.show/ (accessed June 19, 2019).

10 Sam Wineberg et al., Stanford History Education Group, "Evaluating Information: The Cornerstone of Civic Online Reasoning: Executive Summary," Stanford University, November 22, 2016, https://stacks.stanford.edu/file/druid:fv751yt5934/SHEG%20Evaluating%20Information%20Online.pdf.

11 Monica Anderson and JingJing Jiang, "Teens, Social Media, and Technology," Pew Research Center, *Internet & Technology*, May 31, 2018, http://www.pewinternet.org/2018/05/31/teens-social-media-technology-2018/.

12 danah boyd, *It's Complicated: The Social Lives of Networked Teens* (New Haven, CT: Yale University Press, 2014).

13 For information on Soap AI, see Crunchbase, https://www.crunchbase.com/organization/soap-ai (accessed February 20, 2019).

14 Institute for the Future, "Ethical OS: A Guide to Anticipating the Future Impact of Today's Technology," https://ethicalos.org/.

15 Cherri M. Pancake, "Computing's Hippocratic Oath Is Here," *Fast Company*, August 9, 2018, https://www.fastcompany.com/90215922/why-we-spent-two-years-rewriting-the-code-of-ethics-for-computing.

16 Andrew Perrin and JingJing Jiang, "A Quarter of Americans Are Online Almost Constantly," Pew Research Center, *Internet & Technology*, March 14, 2018, http://www.pewresearch.org/fact-tank/2018/03/14/about-a-quarter-of-americans-report-going-online-almost-constantly/.

17 Jacob Poushter, Caldwell Bishop, and Hanyu Chwe, "Social Media Use Continues to Rise in Developing Countries but Plateaus across Developed Ones," Pew Research Center, *Internet & Technology*, June 19, 2018, http://www.

pewglobal.org/2018/06/19/social-media-use-continues-to-rise-in-developing-countries-but-plateaus-across-developed-ones/.

18 Lee Fang, "Google Won't Renew Its Drone AI Contract, but It May Still Sign Future Military AI Contracts," *Intercept*, June 1, 2018, https://theintercept.com/2018/06/01/google-drone-ai-project-maven-contract-renew/.

19 Camila Domonoske, "Google Announces It Will Stop Allowing Ads for Payday Lenders," *The Two-Way*, NPR, May 11, 2016, https://www.npr.org/sections/thetwo-way/2016/05/11/477633475/google-announces-it-will-stop-allowing-ads-for-payday-lenders.

20 Daniel Funke, "Four Major Tech Companies Take New Steps to Combat Fake News," *Poynter*, July 12, 2018, https://www.poynter.org/fact-checking/2018/four-major-tech-companies-take-new-steps-to-combat-fake-news/.

21 Andrew Arsht and Daniel Etcovitch, "The Human Cost of Online Content Moderation," *Harvard Journal of Law and Technology*, March 2, 2018, https://jolt.law.harvard.edu/digest/the-human-cost-of-online-content-moderation.

22 Ravel, Sridharan, and Woolley, "Principles and Policies to Counter Deceptive Digital Politics."

23 Samidh Chakrabarti, "Hard Questions: What Effect Does Social Media Have on Democracy?" *Facebook Newsroom*, January 22, 2018, https://newsroom.fb.com/news/2018/01/effect-social-media-democracy/.

24 Betty Reid Soskin and Luvvie Ajayi, interview at Makers conference, 2018, www.makers.com/videos/5a79ff2d44a64b138fefea2d.

25 Jade Greear, "Speaking Truth to Power," *Huffington Post*, December 22, 2015, https://www.huffpost.com/entry/speaking-truth-to-power_b_8824094.

Acknowledgments

Writing a book is work and don't let anyone tell you differently. Luckily, or maybe unluckily, for me, I had just finished writing my dissertation when I started doing this one. During that last project, I tried every which way to avoid writing. "It's Thursday," I'd tell myself. Or else something like, "I've been traveling for work, I deserve a break and a stiff drink. No need to write today." Eventually, though, I forced myself to write for two hours, from 8:00 a.m. to 10:00 a.m., every day. I stuck with it most days, though certainly not all of them. I'm only human, after all, and not one of the tireless AI bots that generate content endlessly, as you've read about already. I took that two-hour daily writing practice I'd followed while writing my dissertation and applied it to this book project. I still had to spend whole weeks writing here and there, and I had to be fiercely protective of my writing time for a very long period. With this in mind, the first people I want to thank are my wife and best friend, Samantha, and my family: Mum, Dad, Oliver, Justin, Morris, Mathilda, and Basket as well as the Woolley, Donaldson, Shorey, Joens, Bossenger, Loor, and Westlund families. Thanks also to my dear friends: Joe, Ike, Anjuli, Peter, Nick, Basket, Josh, Ayda, Rick, Zach, Doug, Mel, Carleigh, Patrick, and Trent. Thank you for putting up with me spending so much time chained to my desk typing away at the computer. Thank you for putting up with me repeatedly asking you what you thought about different uses of virtual reality or ways of manipulating social media. And thank you especially for supporting me, loving me, and helping me to grow. To my colleagues and collaborators, thank you so much for your perseverance, hard work, and comradery. I would be nothing without the people with whom I've learned to do research. In a way, I grew up to be an actual functioning adult among these people. It certainly took a while. Thank you, in particular but in no particular order, to Nick Monaco, Phil Howard, Lisa Maria Neudert, Samantha Bradshaw, Rob Gorwa, Katie Joseff, Marina Gorbis, Ashley Hemstreet, Matthew Adeiza, Tim Hwang, Doug Guilbeault, Gina Neff, Kirsten Foot, Benjamin Mako Hill, Christine Harold, Ryan Calo, Dan Newman, Ann Ravel, Hamsini Sridharan, danah boyd, Kate Crawford, Camille Francois, Ben Nimmo, Lori McGlinchey, Chancellar Williams, Kelly Born, Anamitra Deb, Joan Donovan, Adam Ellick, Craig Silverman, Karen Kornbluh, Nate TeBluntenhuis, Becca Lewis, Brittan Heller, Daniel Kelly, Talia Stroud, Mike Ananny, Michael McFaul, Nate Persily, Brandie Nonnecke, Stephen Reese, Jay Bernhardt, Kathleen McElroy, and anyone else that I have forgotten here. It is through your mentorship, collaboration, trust, and advice that I am able to do the things I do. Without you, this book would not exist. Thanks also to the European Research Council and the National Science Foundation for supporting early research. Finally, thank you to my editor, Ben Adams, my publisher, PublicAffairs, and my agent, Chris Parris-Lamb. Thanks for taking a chance on me in this game we call reality.

Index

About the Author

SAMANTHA SHOREY

Dr. Samuel Woolley is a writer and researcher specializing in the study of automation/AI, politics, persuasion, and social media. He is an assistant professor in the School of Journalism at the University of Texas at Austin. He is the founding director of the Digital Intelligence Lab at the Institute for the Future, a fifty-year-old think-tank based in the heart of Silicon Valley. Woolley is co-founder and former research director of the Computational Propaganda Project at the Oxford Internet Institute, University of Oxford. He has held research fellowships at the Anti-Defamation League, the German Marshall Fund of the United States, and Google Jigsaw. He has written articles for a variety of publications, including *Wired*, *Atlantic*, *Motherboard*, *TechCrunch*, and *Slate*. For his work, he has been featured in the *New York Times*, the *Washington Post*, and the *Wall Street Journal* and on *The Today Show*, the BBC's *News at Ten*, and NBC's *Nightly News*. His work has been presented to members of NATO, the US Congress, and the UK Parliament. He lives in Austin with his wife, Samantha, and dog, Basket, and tweets from @samuelwoolley.